THE PLANT WORLD

A highly evolved plant, the thistle *(Sonchus arvensis)* buds, flowers, and disperses its seeds simultaneously.

THE WORLD BOOK ENCYCLOPEDIA OF SCIENCE
VOLUME 5

THE PLANT WORLD

WORLD BOOK, INC.
a Scott Fetzer company

CHICAGO

Staff

President
Robert C. Martin

Vice President and Editor in Chief
Michael Ross

Editorial

Managing Editor
Maureen Mostyn Liebenson

Writers
Bonny Davidson
Karen Ingebretsen
Rita Vander Meulen

Permissions Editor
Janet T. Peterson

Indexer
David Pofelski

Executive Director of Research and Product Development
Paul A. Kobasa

Researchers
Lynn Durbin
Cheryl Graham
Karen McCormack
Loranne Shields

Consultant
Lawrence C. Bliss
Professor Emeritus of Botany
University of Washington

Art

Executive Director
Roberta Dimmer

Art Director
Wilma Stevens

Senior Designer
Isaiah Sheppard

Cover Design
Chestnut House

Photography Manager
Sandra Dyrlund

Product Production

Senior Manager, Pre-press and Manufacturing
Carma Fazio

Manager, Manufacturing
Barbara Podczerwinski

Senior Production Manager
Madelyn Underwood

Manufacturing Production Assistant
Valerie Piarowski

Proofreaders
Anne Dillon
Chad Rubel

Text Processing
Curley Hunter
Gwendolyn Johnson

© 2001, 2000, 1997 World Book, Inc. All rights reserved. This volume may not be reproduced in whole or in part in any form without prior written permission from the publisher.

World Book, Inc.
233 N. Michigan Ave.
Chicago, IL 60601

For information on sales to schools and libraries, call 1-800-WORLDBK (967-5325), or visit our Web site at http://www.worldbook.com

© 1989, 1984 Verlagsgruppe Bertelsmann International GmbH, Munich.

Library of Congress Catalog Card No. 00-109505
ISBN: 0-7166-3399-X (set)
ISBN: 0-7166-3354-X (vol. 5)
Printed in the United States of America

4 5 6 7 8 9 06 05 04 03 02 01 00

Contents

Preface 6	Arctic tundra 84
Careers 8	Coniferous forests 86
	Temperate forests 88
The plant kingdom 12	Tropical rain forests 90
Plant classification 14	
Special feature:	*Special feature:*
What is and is not a plant 16	**The vanishing rain forest** 94
Plant structure and photosynthesis 18	Heath and moorland 96
Plant responses 24	Deserts 98
Nonflowering plants 26	Grasslands and savanna 100
Fungi 30	Aquatics 104
Mosses and liverworts 34	Saprophytes and symbionts 108
Club mosses and horsetails 36	**Plant products** 110
Ferns 38	Wood 112
Seedferns and cycads 40	Woodchip 114
Cone-bearing plants 42	Wood pulp and paper 116
Flowering plants 46	Rayon and cellulose 120
Sexual reproduction 48	Vegetable fibers 122
Seed dispersal 52	Alcohol and other chemicals 126
Vegetative propagation 54	Rubber 130
Monocotyledons 56	Latex and other plant gums 132
Dicotyledons 60	**Conservation and reclamation** 134
Herbaceous plants 62	Forest management 138
Shrubs 66	Shore and desert reclamation 142
Climbers 68	**Classification of the plant world** 144
Trees 70	**Glossary** 146
Special adaptations 78	**Index** 153
Swamps and marshes 80	**Credits** 159
Alpine tundra 82	

Preface

The Plant World, like the other volumes in the *Encyclopedia of Science*, deals with a specific subject, in this case, botany. It begins with the various approaches to the science and introduces the basic life processes of plants. The categories of plants from the simplest to the most highly evolved are described, as well as how these plants have adapted to their environmental conditions. The economic uses of the plant kingdom are also included in the text and lead to a final discussion of the exploitation and threatened extinction of plants and the consequent need for their conservation.

The editorial approach

The object of the *Encyclopedia of Science* is to explain for adults and children alike the many aspects of science that are not only fascinating in themselves but are also vitally important for an understanding of the world today. To achieve this, the books in this series are straightforward and concise, accurate in content, and are clearly and attractively presented.

The often forbidding appearance of traditional science publications has been completely avoided in the *Encyclopedia of Science*. Approximately equal proportions of illustrations and text make even the most unfamiliar subjects interesting and attractive. Even more important, all of the drawings have been created specially to complement the text, each explaining a topic that can be difficult to understand through the printed word alone.

The thorough application of these principles has created a publication that covers its subject in an interesting and stimulating way, and that will prove to be an invaluable work of reference and education for many years to come.

The advance of science

One of the most exciting and challenging aspects of science is that its frontiers are constantly being revised and extended, and new developments are occurring all the time. Its advance depends largely on observation, experimentation, and debate, which generate theories that have to be tested and even then stand only until they are replaced by better concepts. For this reason, it is difficult for any science publication to be completely comprehensive. It is possible, however, to provide a thorough foundation that ensures that any such advances can be comprehended. It is the purpose of each book in this series to create such a foundation, by providing all the basic knowledge in the particular area of science it describes.

How to use this book

This book can be used in two basic ways.

The first, and more conventional, way is to start at the beginning and to read through to the end, which gives a coherent and thorough picture of the subject and opens a resource of basic information that can be returned to for re-reading and reference.

The second allows the book to be used as a library of information presented subject by subject, which the reader can consult piece by piece as required.

All articles are prepared and presented so that the subject is equally accessible by either method. Topics are arranged in a logical sequence, outlined in the contents list. The index allows access to more specific points.

Within an article, scientific terms are explained in the main text where an understanding of them is central to the understanding of the subject as a whole. There is also an alphabetical glossary of terms at the end of the book, so that the reader's memory can be refreshed and so that the book can be used for quick reference whenever necessary.

Each volume also contains a section on the various careers that pertain to the volume's subject.

The sample two-page article *(right)* shows the important elements of this editorial plan and illustrates the way in which this organization permits maximum flexibility of use.

(A) **Article title** gives the reader an immediate reference point.

(B) **Section title** quickly shows the reader how information is arranged within the article.

(C) **Main text** consists of narrative information set out in a logical manner, avoiding biographical and technical details that might tend to interrupt the story line and hamper the reader's progress.

(D) **Illustrations** include specially commissioned drawings and diagrams and carefully selected photographs, which expand, clarify, and add to the main text.

(E) **Captions** explain the illustrations and make the connection between the textual and the visual elements of the article.

(F) **Labels** help the reader to identify the parts of the illustrations that are referred to in the captions.

(G) **Theme images,** where appropriate, are included in the top left-hand corner of the left-hand page, to emphasize a central element of information or to create a visual link between different but related articles.

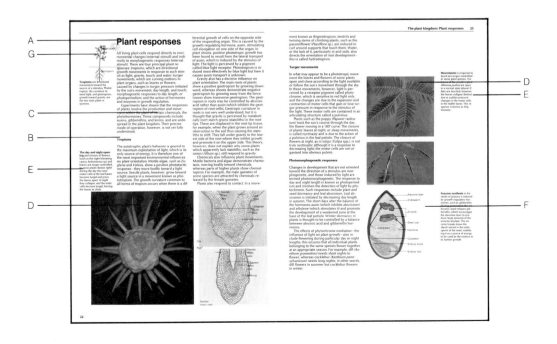

CAREERS

Botanists, *above,* examine lettuce raised indoors by hydroponics.

A**gronomists,** also known as crop research scientists, are botanists who develop new methods of growing field crops for more efficient production, higher yield, and improved quality. Agronomists use the principles of genetics, cellular biology, and physiology to increase and improve the quality and value of crops while preserving the environment and maintaining optimum soil conditions. They conduct breeding studies on farms and at experimental stations to determine the best ways to make plants larger, healthier, and more resistant to disease, weeds, and pests. These botanists also perform extensive soil studies, examining the chemistry, microbiology, mineralogy, and fertility of soil to learn how plants respond to varying soil conditions. Agronomists use their knowledge to improve food crops, such as fruits, vegetables, and grains, as well as other types of crops, such as turf grass and cotton.

A**rboriculturists,** also known as "tree doctors," are trained professionals who care for the trees that line our streets and dot our landscape. Arboriculturists consult with landscape architects to determine which trees to plant in a particular park or building site. To ensure that trees under their care grow healthy and strong, these specialists develop and supervise pruning, fertilizing, and pest-management programs. They fertilize trees to make sure they are getting the proper nutrients to grow healthy and strong and to resist disease. They prune trees to rid them of dead or diseased branches, improve their shape, and promote new growth. If a tree becomes infested with pests, arboriculturists must decide which measures are best to take for the tree and the surrounding environment—whether to add the pest's natural enemies to the area around the tree or to use pesticides.

Arboriculturists spend a great deal of time outdoors, usually working in crews of three or four. In addition to having an extensive knowledge of trees and their environment, arboriculturists must also be familiar with the equipment and methods used in caring for them. These include pruning tools, drills, chain saws, and techniques such as "bracing," which involves installing special supports in tall trees to help prevent branches from being broken during storms and other extreme weather conditions.

B**otanists** are scientists who study all aspects of plants and their environment. Some botanists conduct their investigations and experiments in research laboratories, using sophisticated equipment such as electron microscopes to learn more about the growth processes, anatomy, physiology, and genetic structure of plants. Others, known as field botanists, travel to faraway places to examine plants in their natural habitat.

Botanists have an important role in the search for new and better ways to use plants for food and medicine, and specialties within this field are as diverse as the plant kingdom itself. Botanists who study plant biology focus on the cell and tissue structure of plants, their chemical makeup,

A field botanist, *below,* uses a laser scanner and computer to record the progress of a diseased tree.

and their functions and vital processes. Botanists specializing in the applied plant sciences focus on the practical value of plants and investigate ways to use them for the benefit of people. Some botanists specialize in the study of a particular type of plant, such as mycology (the biology of fungi), pteridology (the study of ferns and similar plants), and lichenology (the biology of lichens). Botanists called taxonomists name and classify plants—and sometimes discover plant species that have never before been identified.

Floriculturists are involved in the production and use of flowers and foliage plants. They work mainly with houseplants, flowers, and greenery for floral arrangements. Floriculturists who work in greenhouses and garden centers keep plants healthy by watering, fertilizing, and inspecting for disease. Floricultural marketing specialists ensure that cut flowers and potted plants are properly packed and shipped from the commercial grower to retail stores. Inventory specialists in the floricultural industry schedule planting cycles to ensure that sufficient flowers are available during times of high demand, such as poinsettias for Christmas and roses for Valentine's Day. Floral designers are floriculturists who design and create floral arrangements. They use their artistic skills and knowledge of flowers to create designs that convey a certain feeling or mood.

Foresters manage and develop forestlands and resources while preserving and protecting the environment. To ensure that forests are being used responsibly for economic and recreational purposes, foresters plan and carry out forestation, cutting, and reforestation programs. To develop these programs, foresters carefully map forest areas, measure standing timber, and estimate future growth. They also advise timber companies on the most effective ways of cutting and removing trees with minimal disturbance to forests and their wildlife. Foresters are also concerned about preventing fires, which are often devastating to forests. These specialists conduct fire-prevention programs and supervise airplane patrols over vast forest areas. If fire breaks out, foresters direct the efforts of fire-fighting crews. Sometimes foresters conduct *prescribed burnings,* in which small fires are deliberately set in the litter of the forest floor to reduce the potential for a large fire.

Horticultural scientists use their knowledge of plant biology, chemistry, and physiology to produce new varieties of plants that are especially beautiful, hardy, and

A horticulturist, *above,* trims plants at a botanical garden.

productive. For example, horticultural scientists experiment with the production of *hybrids* by crossing two plant breeds to create a single plant with the best characteristics of the two parent plants. They also use a technique called *tissue culture* to create new plant species. Horticultural scientists conduct extensive experiments to determine the environment and nutrition necessary for plant growth. They test plants for their adaptability to different climates, soils, uses, and processes. These specialists also work to find ways to control plant diseases and improve plant resistance to pests. They conduct most of their research at agricultural experiment stations, arboretums, botanical gardens, and colleges and universities.

Horticultural therapists plan, manage, and evaluate therapeutic gardening programs to help rehabilitate physically handicapped and mentally handicapped people, substance abusers, criminals, and the socially dis-

Careers 10

Paley Park in New York City was designed by a landscape architect.

Landscape architects plan, design, and manage development projects that use land in a practical way while enhancing the beauty of our natural surroundings. Landscape architects create plans for a wide variety of projects, from botanical gardens to national parks, resorts, playgrounds, residential projects, corporate sites, and city squares. They also design walls, fences, steps, pavement patterns, and planting arrangements.

Landscape architects work on resource-management programs and such projects as wetlands restoration, and play an important role in the preservation of historical landmarks. These specialists are involved in all phases of a land-development project, from land planning to site design. They use their knowledge of climate, water supply, vegetation, and soil composition to ensure that a development project harmonizes with its surrounding environment, while avoiding erosion, flooding, and air and water pollution. They work with architects to fit buildings and other structures into land formations, making the best use of sunlight, breezes, and the beauty of nature.

Medical botanists explore ways to use plants to treat and cure human disease. Since ancient times, people have used plants to relieve pain and reduce or eliminate disease symptoms, and today's medical botanists continue to investigate the many ways in which plants benefit human health. For example, the bark of the cinchona tree provides the drug quinine, which is used in treating malaria. And digitalis, used to treat certain heart diseases, comes from the dried leaves of the purple foxglove, a common garden flower. Medical botanists conduct their research in the field as well as in the laboratory. Because there may be plants in the tropical rain forest with great potential to treat human disease, medical botanists often set up small research stations deep in the jungle. There, they study plant specimens and gather information from local people about using plant species in the immediate surroundings as folk remedies. Medical botanists work for research institutes, academic institutions, agricultural businesses, and pharmaceutical companies.

Plant ecologists are botanists who study the relationship between plants and their environment by investigating their life history, light and soil requirements, and resistance to disease and insects. These scientists collect and analyze information about how plants interact with one another and with animals. Plant ecolo-

advantaged. In horticultural therapy, plants and horticultural activities are used to improve physical, psychological, social, and mental well-being. Activities are designed to enhance self-esteem, encourage a sense of responsibility, and develop motor skills and problem-solving abilities. These include propagating, caring for, and harvesting a variety of plants in indoor and outdoor garden settings. Patients release stress and anger through the physical exertion of planting, watering, and weeding, and they also experience the emotional gratification of nurturing living things. Horticultural therapists are highly educated in the fields of horticulture, psychology, and education, and work side by side with other health-care professionals, including psychiatrists, psychologists, social workers, and occupational therapists.

A medical botanist, *left,* works with a native expert to locate plants for use in medications in the Amazon Basin in Ecuador.

gists also study the adaptability of plant species to new and altered environmental conditions, such as changes in soil type, climate, and altitude. They work with government agencies and industry to discover the effect of natural events and human activities on plant life and the environment. For example, if a hardwood forest is damaged by a hurricane, a plant ecologist would determine how long it will take the forest to recover. A plant ecologist may also work along riverbanks, studying the effects of water pollution on wetland plants. Although plant ecologists conduct much of their research outdoors, they also spend time in laboratories, libraries, and at their computer terminals learning more about the natural world.

Plant pathologists identify plant diseases and experiment with various treatments and cures—a vitally important study, since all life on our planet depends on plants. For example, outbreaks of plant disease can cause major damage to crops or destroy an important species. Plant pathologists are highly trained scientists with an extensive knowledge of biochemistry, ecology, botany, epidemiology, genetics, microbiology, and physiology. They use their knowledge to identify the organisms and environmental conditions that cause disease in plants. They also study how diseases affect the growth, yield, or quality of a plant product. Plant pathologists conduct their investigations in the field and in the laboratory, always searching for new ways to use chemical, biological, and cultural techniques to control plant disease. They work for universities, chemical and pesticide companies, and government agencies.

Soil conservationists plan and manage programs for responsible land use, soil erosion control, and water conservation. These scientists use their knowledge of soil fertility, microbiology, physics, and chemistry to develop effective soil-management practices, such as crop rotation, reforestation, contour plowing, and terracing. Their programs allow for the most productive use of the land without damaging or destroying the health and fertility of the soil. In field investigations, especially land mapping, soil conservationists use a special instrument called a stereoscope to create a three-dimensional picture of the landscape. In this way, they gather important information about distinctive features of the land, such as moisture level, vegetation, and topography. Soil conservationists work closely with government agencies, farmers, foresters, miners, and rural and urban planners.

Soil conservationists help plan strip cropping to reduce soil erosion on sloping land.

The plant kingdom

Of the more advanced living organisms on earth, the plants evolved before the animals. The oldest plant fossils date from the Precambrian era, over 3 billion years ago. It was not until the Cambrian period almost 600 million years ago, however, that animals evolved, by which time plants had prepared the conditions in which animal life could exist. The green plants manufactured sugar by photosynthesis and in doing so released oxygen, which revolutionized the atmosphere of the planet. Only then could the air be utilized by animals.

Since the first appearance on earth of plant life, hundreds of thousands of plant species have populated the planet, many of which are now extinct. Today about 400,000 species of plants are known, although the true number is probably considerably more, because new species are constantly being discovered and identified. Traditionally, the plants are divided into the lower, or nonflowering plants, and those that produce flowers and seeds.

What is a plant?

Plant cell structure usually distinguishes these organisms from animals. The cell has a rigid, cellulose wall, rather than a flexible membrane. Large, permanent vacuoles occur in the center of the cytoplasm, whereas animal cells have small temporary ones, or none at all. Plant cells have chloroplasts that contain chlorophyll; animal cells do not.

A plant is generally considered to be a living organism that manufactures its own food (it is autotrophic). There are some, however, that are parasitic, and others that are saprophytes, processing food outside their bodies and then absorbing it. Plants are usually immobile, functioning on the spot.

Even so, there is some overlap between lower animals and simple plants, and the distinction between them is not always clear. Many unicellular and some multicellular algae move around like animals, but are autotrophic; some surround and ingest their food (they are phagotrophic) like protozoa, but do not have a true cell wall, only a membrane; slime molds are also phagotrophic but have a cell membrane during their feeding stage only and, unlike protozoa, reproduce by spores (they are sporophytic).

The most primitive lower plants have no root structures, conducting or supporting tissues. Vascular plants, in contrast, have a system of vessels that transport water and nutrients. These higher plants are differentiated into stems, leaves, roots and flowers; they also have supporting tissues. Most lower plants are spore-producing, or sporophytic, whereas the higher plants are seed-producing, or spermatophytic.

Lower plants

The simplest organisms are the bacteria and cyanobacteria, which are called prokaryotes because they have no distinct nucleus in their cells. All other plants are called eukaryotes—that is, each cell has a true nucleus and more complex organelles. Bacteria are single-celled but contain no chlorophyll and, rather than surviving by photosynthesis, use a variety of biochemical means to obtain energy. Some manufacture their energy supplies from chemicals in their immediate environment whereas others rely on finding it ready-made. These bacteria include the disease-forming types

Animal cells *(below)* and plant cells *(below right)* differ in several respects but also contain some of the same organelles. The main distinctions between them are that the animal membrane is plastic while that of the plant is rigid and contains cellulose; plant cells have large sap vacuoles, but animal cells have none; chloroplasts containing chlorophyll are present in many plant cells, but not in animal cells.

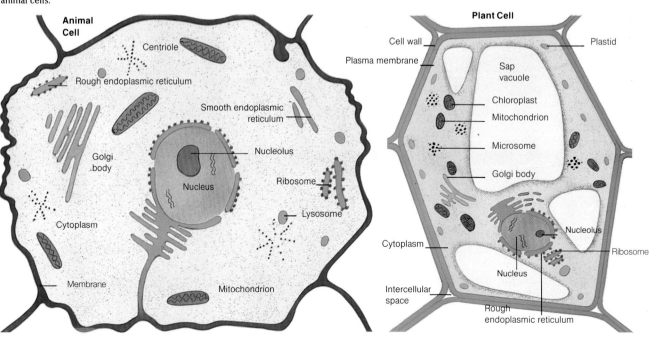

that inhabit the bodies of humans and other animals. They reproduce simply by splitting in two (binary fission) and, occasionally, by spores. Bacteria comprise about 4,000 species.

The cyanobacteria (blue-green algae) are grouped with the bacteria because they share some structural features: for example, their DNA is not arranged into chromosomes; and they do not have a nuclear membrane. Some of the other algae are more closely related to the higher plants in their cell structure. Many, such as *Euglena,* consist of single cells capable of independent existence. Others are composed of chains (filaments) of cells, and the large seaweeds are made up of thousands of cells. They also vary widely in their pigmentation, cell wall composition, and the kind of foodstuff they store. Algal reproduction may be asexual—by binary fission—or sexual, by gametes. There are approximately 21,000 species of algae.

In common with the bacteria, the fungi do not contain chlorophyll and, therefore, do not photosynthesize. Instead, they live either as parasites on other plants or animals, obtaining food ready-processed or as saprophytes. Like the bacteria, they function as decomposers, releasing nutrients that support higher organisms. Fungi include molds, yeasts, rusts, smuts, mushrooms, and toadstools. This group comprises more than 100,000 species. Most reproduce sexually and asexually, the typical mushroom-shaped fruiting body producing spores by sexual means.

The lichens are compound organisms of fungi and algae, which have a symbiotic relationship. This successful group is autotrophic and survives in almost every habitat; it contains about 18,000 species.

Mosses and liverworts (the bryophytes) are represented by more or less the same number of species as lichens are. They photosynthesize but are nevertheless regarded as primitive land-dwelling plants because they do not have roots or a truly differentiated vascular system. These green plants typically have an alternation of sexual and asexual generations (gametophyte and sporophyte respectively).

The ferns, club mosses, and horsetails—the pteridophytes—do have a vascular system. Like the mosses, these plants also display alternation of generations, but the mature plant (the sporophyte) is much more complex than that of the bryophytes, with differentiated growing and conducting tissues. Therefore, they can grow much larger in size. There are about 12,000 known species of pteridophyte, which are distributed worldwide; this is, however, a tiny fraction of the number that dominated the earth about 400 million years ago.

Higher plants

The higher plants are distinguished from the lower ones by the development of the seed and with it the restriction of the gametophyte (sexual) generation to a fusion of cells within the sporophyte. This development reduced the risks of the reproductive process and (except in primitive forms such as cycads and ginkgos) lessened the need for water as an essential medium for the completion of fertilization.

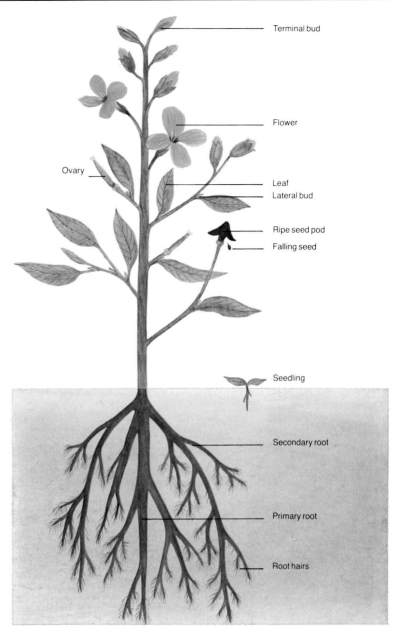

About 250,000 species of seed-producing plants are recognized today. Of these some 500 species are gymnosperms (conifers and cycads). Typically, the seeds of gymnosperms are naked, compared with those of angiosperms, whose seeds are enclosed in a tissue coating called an ovary. The angiosperms, or flowering plants, now dominate the earth's vegetation, having adapted to survive in almost every habitat.

The basic structures of many flowering plants include a primary root, which bears secondary roots; a main, woody stem; lateral branches from which grow photosynthetic leaves; and flowers. Lateral buds in the angles between the branches and the leaves, or the main stem, are suppressed regions of growth, which may later develop into branches themselves. After blooming, when fertilization occurs, the flower petals drop off, leaving the ovary in which the seeds develop. The seeds are shed when they are ripe and later grow into seedlings.

Plant classification

Field rose
(*Rosa arvensis*)

Kingdom
Plantae
Division
Magnoliophyta
Class
Magnoliopsida
Subclass
Rosidae
Order
Rosales
Family
Rosaceae
Genus
Rosa
Species
arvensis

Scientific classification is the system most commonly used in describing plants. Using this method, a common field rose could first be grouped into those organisms that are eukaryotes—that is, they have nuclei in their cells. This group embraces most animals and plants and is considered to be a kingdom by some botanists. The grouping can then be broken down in stages to the basic unit—the plant's species name, *Rosa arvensis*.

The classification of plants and animals provides a logical system by which living organisms can be accurately identified. But in its widest sense, plant classification means more than just giving a name to each of the more than 400,000 or so species of which the plant kingdom is composed. It consists of grouping plants in categories of ever-increasing size—a process known as scientific classification. If a classification also takes account of the relationships that exist between plants then it is described as systematic. The entire process of classifying organisms, and the study of classifications past and present, is embraced by the blanket term taxonomy.

The foundations of modern taxonomy were laid in the mid-eighteenth century with the work of the Swedish naturalist Carl von Linné (1707-1778), better known by the Latinized version of his name—Linnaeus. It was Linnaeus who, with the publication of his *Species Plantarum* in 1753, established the system of binomial nomenclature for plants—such as *Rosa arvensis* for the field rose.

Naming and grouping

Today the naming of plants and their placement in a standard, accepted hierarchical system of classification is governed by strict rules laid down in *The International Code of Botanical Nomenclature* (ICBN). This rule book recognizes 12 principal ranks of diminishing size, ranging from kingdom through phylum and species to form, with subcategories that can be designated for each category. Of these, the basic "unit" of classification is generally considered to be the species.

A species is a group of individuals that are so similar as to be more or less identical and that can interbreed. Similar species are grouped into genera, similar genera into families, and so on.

Despite the fact that plants are grouped and named in such a rigid way, they are not classified using just one system. Scientific classification is an interpretation of facts. It is based on the opinion and judgment a biologist forms after studying specimens of animals and plants. Most biologists use the same basic framework for classification. But not all biologists agree on how individual animals and plants fit into this scheme, and so classifications often differ in details. The system used here differs slightly from that discussed in *The World Book Encyclopedia*. For more information on scientific classification, see the *World Book* articles CLASSIFICATION, SCIENTIFIC and PLANT (A classification of the plant kingdom).

Any system of classification that takes into account only one or a few plant (or animal) characteristics is described as artificial. Such an artificial system could group plants, for example, according to their color, petal number, or edibility. While they are useful for specific

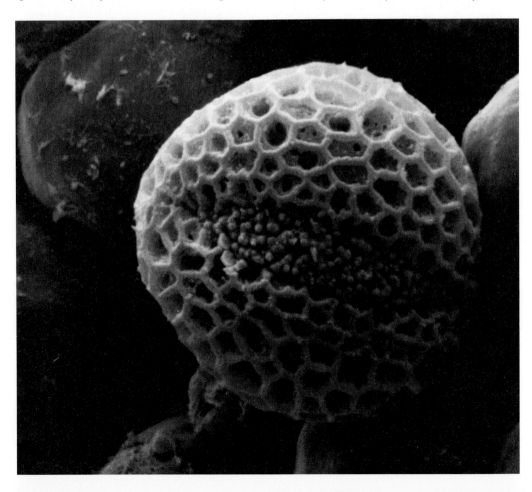

Scanning electron micrographs are often used to study pollen grains and spores. Such microscopic examination of these bodies reveals their structural characteristics, which can be used as a means of grouping together species of plants.

The plant kingdom: Plant classification

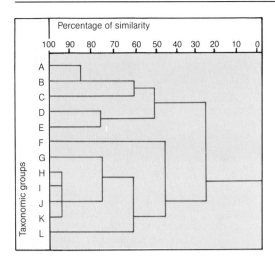
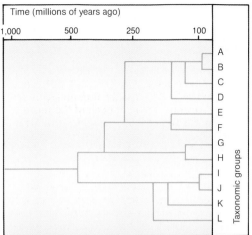

Dendrograms are a graphic means of showing the relationships between groups of plants. A phenogram *(far left)* charts the percentage of phenotypic similarities between a number of taxonomic groups. A cladogram *(left)* indicates the evolutionary pathways of species over the past millions of years to reveal their lineage relationships.

purposes, such systems do not have great worth in the wider botanical context. By comparison natural systems, based on the analysis of a great many characteristics shared (or not shared) by plants, are very useful.

In some aspects of botanical study, however, demands are made that are not necessarily met by a natural system of classification; notable among these is the study of the evolutionary relationships between plants. This system is designed to reflect in particular the evolutionary pathways along which plants have progressed over millions of years of development.

Gathering information

Any system of classification relies on the accumulation of vast amounts of data about plants. This information includes the gross structure of a plant (its morphology), often with special reference to flowers and other organs involved in reproduction. Plants are measured and examined microscopically in every detail with light and electron microscopes. The development of each plant is studied macro- and microscopically, and its number of chromosomes, breeding systems, geographical distribution, and ecology are all considered. Chemistry and DNA sequencing can be significant, and the fossil record can also provide valuable insight into the way in which plants should be grouped together.

The practice of numerical taxonomy is based on the construction of similarity tables or data matrices between species, genera, and other classifications. It relies on computers to store and analyze information and to calculate the degrees of similarity between plants. By this means, plants are evaluated and grouped according to the degree of similarity they possess.

Plant classification is a means by which we can keep track of the changing nature of the plant kingdom, for example by monitoring plants that are in danger of extinction, particularly now that many of our traditional natural resources are in short supply or running out. At the laboratory level, experimental taxonomy, which is designed to discover more about the nature and genetic basis of the diversity of species, could hold the key to the preservation of that diversity and its use in such tasks as the improvement of crop plants.

Fossils of plant organs such as leaves are an important source of material for the evolutionary classification of plants.

Illustrated herbals played a significant role in the early description and subsequent classification of plants. Many portrayed the whole plant and its individual parts.

WHAT IS AND IS NOT
a plant

From tiny blades of grass to giant sequoia trees, the thousands of plants in the world enrich our lives. They not only provide us with food and oxygen, but also bring us beauty, shelter, warmth, and clothing. So what makes a plant a plant? Sometimes it's obvious and sometimes it's obscure—and often there's a lot more there than meets the eye. Getting to the bottom of what defines a plant as a plant is like investigating a mystery.

Aristotle divided all living things into two kingdoms—plants and animals. His system of classification—the simplest and the longest-used—is the one used in this encyclopedia. As discussed in the preceding article, this classification was used by Linnaeus, who based his classification on visible structures. Even in this system, however, classifying an organism such as euglena, for example, was confusing because the euglena displays traits characteristic of both kingdoms. Since the early 1900's, and increasingly in recent years, *taxonomists* and *systema-* *tists* have faced the challenge of developing new systems of classification. Taxonomists are scientists who identify, name, and classify organisms, and systematists are scientists who classify organisms or groups of organisms on the basis of their evolutionary relationship.

In the 1950's, formal views of classification slowly began to change. Advances in the study of embryology, molecular biology, and biochemistry presented new and different ways of looking at plants and animals. However, proposals of new kingdoms met with resistance

In the five-kingdom system of classification, a tickseed flower *(Coreopsis tinctoria)*, *(inset)*, is a plant but diatoms, single-celled algae, are not.

until sophisticated technological advancements in microscopy allowed scientists to study the internal structures of both small and large living things. For the first time, the differences between bacteria and fungi were seen as greater than the differences between plants and animals.

New classification systems

In the following years, many new multikingdom systems were proposed, including a 13-kingdom system. The most widely supported of these proposals is the 5-kingdom system presented by ecologist Robert H. Whittaker of Cornell University in 1959.

The five kingdoms of Whittaker's system—Monera, Protista, Fungi, Plantae, and Animalia—are based on microscopic structural and biochemical characteristics as well as on evolutionary relationships. In the five-kingdom system, bacteria were distinguished from plants by being *prokaryotic organisms* (organisms whose cells lack a membrane around the nucleus) and were moved from the plant kingdom into their own kingdom—Monera. Algae (except for blue-green algae, which are Monerans) were moved from the plant kingdom into the kingdom Protista, which they share with protozoa, former members of the animal kingdom. Protista are recognized by what they *don't* do and what they are *not*. For example, unlike the organisms of the plant kingdom, algae do not develop from an embryo contained in a seed. Fungi also have their own kingdom, and differ from other large *eukaryotic organisms* (plants and animals) in that they develop without an embryo.

What is a plant?

After removing bacteria, algae, and fungi, what is left in the kingdom Plantae? The plant kingdom has thousands of species, including mosses and liverworts, club mosses and horsetails, ferns, cone-bearing plants, and flowering plants. Most plants produce their own food through photosynthesis. But technically, plants differ from the other kingdoms based on their life cycle—and not on photosynthesis. All members of the plant kingdom have a life cycle called "alternation of generations." In this type of life cycle, a haploid stage with only one set of chromosomes alternates with a diploid stage that has two sets of chromosomes. Haploid plants are called gametophytes; diploid plants are called sporophytes. In the flowering and cone-bearing plants, only the sporophyte can be seen with the unaided eye. In other plants, such as some mosses, the familiar plant is the gametophyte.

A plant's life cycle is the most basic qualification for membership in the plant kingdom, but there are other criteria. Almost all plants are land dwellers. In addition, unlike fungi, which absorb nutrition, and unlike animals, which eat food, most plants produce their own food through photosynthesis. Kingdom Plantae members are multicellular with specialized cells and tissues and have plastids, chlorophyll, and cellulosic cell walls. Some species of algae, such as the seaweeds, resemble plants and share some plant characteristics. Some of these species alternate generations, have plastids, and have cellulose in their cell walls. They also use chlorophyll to photosynthesize and produce their own food. But they are not considered truly multicellular. Their specialized cells and tissues are much simpler than the complex structures in the true plants, so they are no longer considered plants under the five-kingdom system.

All systems of classification—new and old alike—are created by humans for human use and subject to judgment calls by people whose viewpoints and opinions frequently differ. Further advances in technology, along with the accumulation of knowledge, are certain to bring changes in the classification of organisms. For example, DNA and RNA sequencing have begun to clarify the evolutionary history of bacteria, and may someday lead to reorganization of Moneran classification. But, in the study of plants, it is important to remember always that "a rose by any other name would smell as sweet."

Small red gill mushrooms, *(below)*, are a member of the Fungi kingdom, in the five-kingdom system.

Plant structure and photosynthesis

Two factors that make plants as living organisms different from animals are their structure and their method of obtaining the energy to "power" their life processes. Structurally most plants are based largely on cellulose, a natural polymer made up of atoms of carbon, hydrogen, and oxygen (cellulose is a complex carbohydrate). Plants derive their energy from sunlight, which they use in the process called photosynthesis to bring about the chemical combination of carbon dioxide (from the air) and water (from the soil) to make simple sugars. In addition, for healthy growth plants need various other elements—particularly nitrogen and phosphorus—which they must also obtain from the soil. Transport of the various elements and compounds through the tissues of the plant is carried out by water, and so water plays a key role both in photosynthesis and in running the biochemical factory of which every plant consists. Water is the chief constituent of the fluid within the plant cells that make up the various tissues and structures of a plant.

Cell structure

In 1665, when the British scientist Robert Hooke first turned his simple microscope onto a thin slice of cork (the outermost layer of a woody plant stem), he described what he saw as resembling a collection of "little boxes." Later he coined the word "cell" for the units of plant tissue (later to be applied to animal tissue as well).

Plant cells are "packets" of living protoplasm surrounded by a dead cell wall composed mostly of cellulose, a carbohydrate manufactured by a plant cell's cytoplasm. Cellulose fibers give strength to plant stems, roots, and leaves. It is the substance that makes plant stems stiff.

The cell wall surrounds a central mass of living cytoplasm, which in turn encloses a large vacuole containing cell sap, a mixture of various chemicals in a watery solution. The cytoplasm, which is bounded by a thin cell membrane, is not uniform in appearance or texture but contains various structures called organelles. These include mitochondria, the cell's energy-producers; chloroplasts, which conduct the process of photosynthesis; and lysosomes, which contain digestive enzymes to break down any substances that enter the cell. Also in the cytoplasm is the many-folded endoplasmic reticulum, which produces fats and proteins, and the nucleus, the center for nucleic acids and the genetic material of the cell.

The cells of a plant begin as small units produced at growing points, called meristems. Those at the growing tips of shoots or roots are known as apical meristems; those in the body of a plant include cambium, which produces vascular tissue, and cork cambium, which produces the outer tissue called cork.

Plant tissues

As cells grow they differentiate into the various tissues of the plant. For instance the outermost covering of cells known as the epidermis forms a continuous layer around the plant. The characteristics of epidermal cells depend on their position on the plant, but most of them are flat, platelike cells. Modified versions include the guard cells of stomata (the "breathing holes" in a leaf) and the fine hairs on roots and rootlets. Epidermal cells secrete a waxy cuticle onto their surface and have cell walls containing a great deal of a fatty water-repellent substance called cutin, both of which help to protect the plant.

In older stems and roots the epidermis is frequently replaced by the cork, which is bet-

All green plants, from the mosses on the fallen log to the lush jungle vegetation and the huge buttress-rooted tree, build their structures from chemical substances derived ultimately from the products of photosynthesis. Thus sunlight is the source of energy for the growth of all green plants.

ter at withstanding damage. It contains three different kinds of cells: the cork cambium (phellogen), a typical meristematic tissue that divides to produce cork cells (phellem) to the outside and cortical cells (phelloderm) to the inside. The cork cells include a layer of water-repellent suberin.

The water-conducting tissues (the xylem) are long tubes running the length of the plant and form part of the vascular bundle in angiosperms. There are two types—the tracheids and the vessels—with walls that characteristically show spiral, annular, scalariform (ladder-like), or reticulate (netlike) thickening. Tracheids are long narrow cells about .5 to 3 inches (1 to 7.5 centimeters) long and 0.04 to 0.06 millimeter across, which join at their ends with a very slanting wall. Vessel cells are small but some of the walls between them break down to form a long tube; those walls that remain are perforated by pores. The vessels may be as narrow as only 0.006 millimeter across or up to 0.7 millimeter wide, but in trees they may be as long as 16 feet (5 meters).

The conducting tissues—the xylem and phloem—are produced by a layer of meristematic tissue called the cambium. This vascular tissue is formed primarily in discrete bundles, but the cambium may extend to form a complete ring of tissue within the plant stem. Cells produced by the cambium differentiate into xylem on the inner side and phloem on the outer side. Both occur throughout the plant.

Xylem's main function concerns the movement of water around the plant (by the tracheids and vessels), although it is also a supporting tissue (by means of sclereid and fiber cells) and has some storage functions (using its parenchyma cells). Phloem is concerned with the movement and storage of food, and it also acts as a supporting tissue. Food movement is achieved by sieve tubes. The surrounding tissue, the parenchyma, is again the food store, and phloem fibers are the supporting elements.

Parenchyma consists of living cells capable of growth and division. They have various shapes from stellate (star-shaped in cross section) to the more common polyhedral conformation. Some have thickened walls, and some may be specialized for secretion or excretion. Their general functions are in photosynthesis, storage, and repair. The type specialized as a supporting tissue—with unevenly thickened walls—is called collenchyma. It occurs as strands or bundles in stems, leafstalks, and leaf veins.

Sclereids and fibers occurring as mechanical tissue in both xylem and phloem are examples of sclerenchyma. To fulfill their function as support and strengthening cells they have thickened and often lignified walls. The two types vary in shape: fibers are long and thin, whereas sclereids may vary between polyhedral and elongated and are sometimes branched.

Water conduction

Most herbaceous plants contain from 70 to 95 per cent water, and some algae may contain as much as 98 per cent. The water has a number of functions. One is to act as a hydrostatic skel-

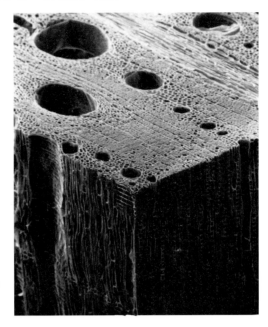

The water-conducting vessels, consisting of joined-up long narrow cells, can be clearly seen in the scanning electron micrograph of a piece of wood (above left). Xylem vessels may have various structures, with the walls thickened in a characteristic way (above). Common forms, shown here in longitudinal section, are spiral (A), annular (B), and the ladderlike scalariform (C). Like the blood vessels in an animal, the vessels become increasingly finer toward a plant's extremities, until eventually they form the threadlike veins of the leaves (left).

eton that helps to support the plant (most plants deprived of water soon wilt). Another role of water is as a medium in which biochemical processes take place. But one of its most important functions is to transport various materials—dissolved gases, minerals, and nutrients—around the plant's tissues.

The continual movement of water through a plant is known as the transpiration stream. Water from the soil enters root hairs by osmosis (that is, it crosses the semipermeable membranes surrounding the cells) and thus dilutes the cell contents. This causes water to flow through the membranes into the next cell, and an osmotic chain is set up through the cells to the water-conducting tissues. A similar chain is set up in the leaves. Cells surrounding the air spaces lose water by evaporation, which concentrates the cell sap. Water from surrounding cells passes in, establishing a chain back to the water-conducting tissues.

The transpiration stream therefore comprises the movement of water into and through the roots, along the xylem vessels, and through the leaf cells to be evaporated or transpired through the stomata of the leaves.

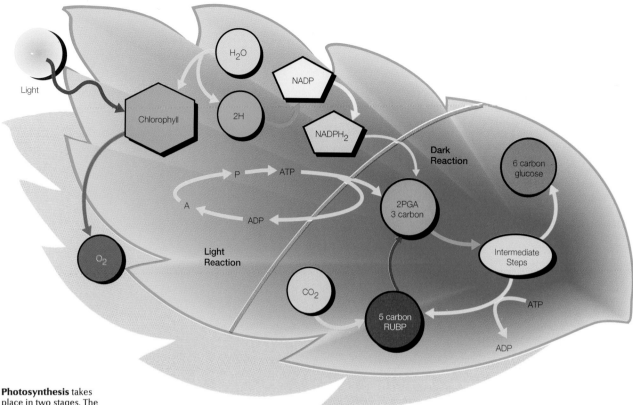

Photosynthesis takes place in two stages. The overall outcome of the light reaction is the splitting of water molecules—using the energy of sunlight absorbed by chlorophyll—to release oxygen and generate energy-rich compounds such as ATP. A cycle of reactions (the Calvin cycle) in the dark reaction combines carbon dioxide with RUBP, hydrogen from $NADPH_2$, to form the sugar glucose.

The closely packed layers called thylakoids, or granae, are the sites within chloroplasts in which the light reaction of photosynthesis takes place.

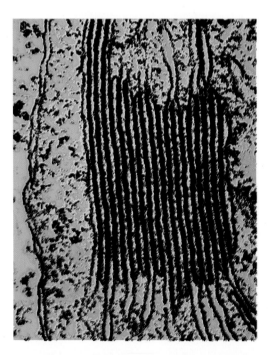

The water in the conducting tissues passes up the plant partly because of the osmotic "root pressure," partly because of the capillary effect of the narrow tubes, and partly because of the osmotic "suction pressure" from the leaves. The rate of movement in the conducting tissues varies, but rates of 10 inches and 13 feet (25 and 400 centimeters) per hour have been recorded. The amount of water loss per plant also varies. A corn plant *(Zea mays)*, for example, may lose nearly a gallon (3.8 liters) of water per day, or approximately 325,000 gallons of water per acre (3,040,000 liters per hectare) in a growing season.

Photosynthesis

Apart from a few bacteria, plants are the only living organisms that are capable of manufacturing their own food. It is the energy of sunlight that powers this process, which is appropriately called photosynthesis. Animals cannot make their own food but rely—directly or indirectly—on plants to produce it.

This is the fundamental difference between plants and animals. All other differences follow on. Since an animal cannot make its own food it must go out and find it—hence it must be mobile. Animals must be able to recognize food when they meet it—hence the sensory and nervous systems that plants lack. Movement entails cells that are flexible—not stiffened with strong walls of cellulose, as plant cells are. Since animals eat plants, or eat other animals that eat plants, it can be said that nearly all the biosphere derives its energy directly from the sun. The only exceptions are in volcanic areas in the dark ocean depths, where the energy to power life is obtained from the erupting chemicals.

The raw materials for photosynthesis are carbon dioxide from the air (taken in through the leaves) and water, usually from the soil (taken in through the roots). The two combine initially to produce simple sugars and oxygen. Sunlight is the energy source, and the green pigment chlorophyll is the means whereby the sunlight can be used. In biochemical terms, the whole process is the reduction of carbon dioxide (to simple sugars) by hydrogen obtained from the breakdown of water mediated by chlorophyll. Photosynthesis takes place as a large number of interrelated stages, but can be summed up overall by the simple equation:

$$6CO_2 + 12H_2O + (light + chlorophyll) \to C_6H_{12}O_6 + 6O_2 + 6H_2O$$

The green color of plants is due to the presence of pigments in the chloroplasts. Chief of these pigments is chlorophyll a, which is found in all green plants and converts light energy in the red and blue wavelengths into chemical form. Most plants also contain other pigments—chlorophyll b, xanthophyll, carotene, and pheophytin—each of which absorbs light of different wavelengths. Light energy in the presence of chlorophyll splits the water molecule. This releases hydrogen (H), which combines with the hydrogen-carrier (NADP), converting it to $NADPH_2$. The energy released (electrons) is transferred to the phosphorus (P) carrier, converting ADP (adenosine diphosphate) to ATP (adenosine triphosphate). The oxygen split from water is released to the atmosphere through the stomates. This entire process, which releases energy stored in ATP and the transfer of hydrogen to $NADPH_2$, is called the "light reaction."

The next state, the "dark reaction," is so called because it does not require light. This cyclic process, called the Calvin cycle, combines the energy of ATP with $NADPH_2$ and carbon dioxide to form a simple carbohydrate (glucose). The cycle begins when a carbon dioxide (CO_2) molecule combines with a molecule of 5-carbon sugar (RUBP-ribulose bisphosphate), which then splits into two molecules of a 3-carbon compound, PGA (phosphoglyceric acid). These two 3-carbon PGA molecules combine with the transfer of hydrogen (H) from $NADPH_2$ and the transfer of the electron energy from ATP to form the simple 6-carbon sugar, glucose.

In many tropical grasses and plants that grow in arid areas the cycle is modified so that the first photosynthetic product is not a 3-carbon molecule but a 4-carbon one (a dicarboxylic acid). This alternative method, the so-called C4 pathway, is more efficient than the C3 pathway.

The rate at which photosynthesis takes place is affected by various external factors. The most obvious of these is the availability of light. At low light levels the rate of photosynthesis is proportional to the light intensity. At higher light levels, a point of light saturation is reached beyond which any increase in light does not increase the rate of photosynthesis. Then other factors become limiting. One of these limiting factors is the carbon dioxide concentration of the air; an increase in carbon dioxide speeds up photosynthesis. Temperature is a limiting factor only to the dark reaction, because it is a chemical process (the light reaction is photochemical, and therefore largely independent of temperature). Water is a limiting factor in that lack of it causes the stomata of the leaves to close, thereby preventing carbon dioxide from entering.

Plant respiration

Photosynthesis effectively traps the energy of sunlight and stores it (as food materials) for fu-

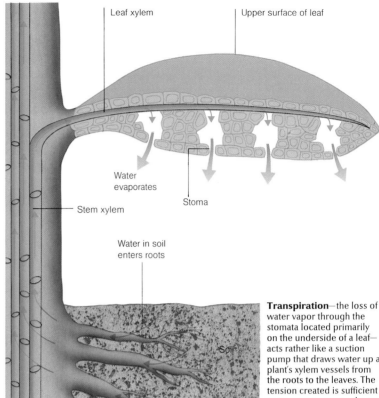

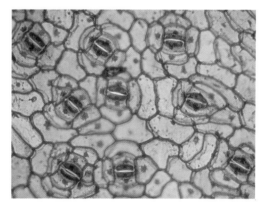

Transpiration—the loss of water vapor through the stomata located primarily on the underside of a leaf—acts rather like a suction pump that draws water up a plant's xylem vessels from the roots to the leaves. The tension created is sufficient to transport water to the top of the tallest tree. Stomata (shown close-up, *below left*) are pores, each one edged with two guard cells that enable them to open and close, depending on external conditions such as humidity and temperature.

ture use. The stored energy is released in the process of respiration, which chemically is an oxidation reaction in which sugars are converted to carbon dioxide and water, with the release of energy. Like photosynthesis, respiration is a stepwise process that can be summed up by the simple equation:

$$C_6H_{12}O_2 + 6O_2 \xrightarrow{enzymes} 6CO_2 + 6H_2O + energy$$

which is the exact opposite of the photosynthesis equation. At a certain light level, the rates of photosynthesis and respiration are in balance. This is known as the light compensation point.

Respiration takes place in the cytoplasm of the cell, principally in the mitochondria. The major sequence of reactions follows a cyclic path known as the Krebs cycle (or citric acid cycle), which is preceded by a shorter metabolic pathway called glycolysis. Using glucose as an example, the first stages involve the splitting of the 6-carbon glucose molecule (by hydrolysis) in a series of reactions to produce pyruvic acid. Next, carbon dioxide is lost from

The plant kingdom: Plant structure and photosynthesis

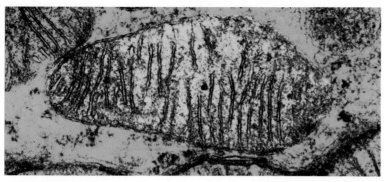

A mitochondrion *(yellow)* is the microscopic body most concerned with cell respiration, as explained in the diagram below.

Cell respiration is a complex process that begins with glycolysis, the breakdown of glucose. The 2-carbon compound acetyl coenzyme A is produced, which then initiates a series of reactions within a mitochondrion. In this series, called the Krebs cycle, more carbon dioxide is released and energy stored in the form of ATP (adenosine triphosphate), 12 molecules of which are produced during each cycle.

the two 3-carbon pyruvic acid molecules, and the resultant 2-carbon acetyl group combines with a carrier enzyme to form acetyl coenzyme A. Then at the start of the Krebs cycle the acetyl group is transferred to a 4-carbon molecule (oxaloacetic acid) to produce the 6-carbon citric acid. Finally this acid is broken down in a series of steps to regenerate the 4-carbon molecule, oxaloacetic acid. Some steps involve the release of carbon dioxide and hydrogen, which is passed via a hydrogen acceptor such as NAD. Energy is released during hydrogen transfer, and some of it converts ADP into ATP.

Some organisms such as yeasts and bacteria can derive their energy in the absence of oxygen, by anaerobic respiration or fermentation, from glycolysis. The route is pyruvic acid to carbon dioxide and acetaldehyde, which is converted to ethyl alcohol (as in winemaking), which in turn may be oxidized to acetic acid. Only a small amount of energy is released, and some of the products are toxic. When oxygen is in short supply—for example, in waterlogged roots—plants may be forced to use anaerobic respiration. Some have developed systems to convert the poisonous product alcohol into less toxic substances (such as organic acids) so that they can employ anaerobic respiration to get through difficult times.

Synthesis of complex compounds

The energy released by respiration is used by plants for various purposes, including the build-up of complex compounds such as carbohydrates, fats, and proteins and to power energy-consuming processes that take place within the cells. A simple sugar, such as glucose, is a basic end product of photosynthesis and is the starting material for a large number of different chemicals. Chemically, glucose is a monosaccharide whose molecules contain six carbon atoms, five of which together with an oxygen atom are arranged in a ring—it is termed a hexose. Each carbon has a hydrogen atom and a hydroxyl group attached to it. It is the replacement of the attached groups or one of the carbon atoms that alters the chemistry of the molecule and creates new compounds. Other important plant monosaccharides include 3-carbon (triose) and 5-carbon (pentose) sugars.

When two hexose molecules join together they form a disaccharide such as sucrose (table sugar), which is formed and stored in plants such as sugar beet and sugar cane. When large numbers of hexose molecules join together, a polysaccharide results. Polysaccharides may be structural, such as the natural polymer—and valuable raw material—cellulose, formed from glucose units. Other structural polysaccharides include mannan from yeast (mannose units), xylan (xylose units), and pectin, materials that form cell walls.

Other groups of polysaccharides have nutrient functions, acting as food stores in plants. Among the most important are starches and inulins. Starch in a storage organ such as a potato tuber, for example, occurs as grains containing two different molecules, amylose and amylopectin. Amylose consists of long chains

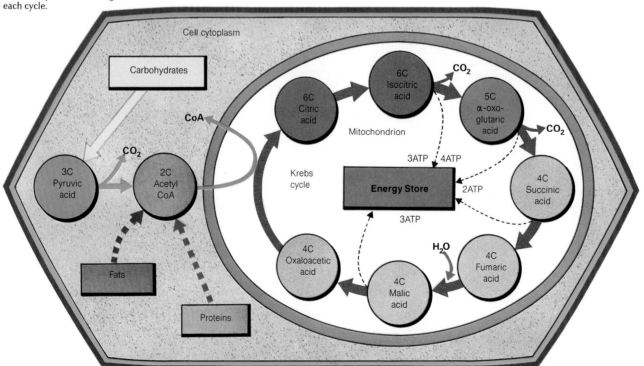

of glucose units, whereas amylopectin has a very branched structure. Inulin is made up mainly of fructose units, with some glucose.

Many seeds have fats and oils as storage material rather than carbohydrates such as saccharides and polysaccharides. The fats are esters of a high molecular weight fatty acid (such as stearic or oleic acid) and an alcohol (usually glycerol). Many plant oils are of economic importance: examples include the oils from olives and groundnuts.

The living parts of cell protoplasm consist of proteins, rather complex molecules containing nitrogen and built up from amino acids. There are 20 naturally occurring amino acids in proteins, and any given protein depends on the number and arrangement of some of these. Important proteins include enzymes, the biological catalysts that bring about and control most of the chemical processes in the cells; some work only in conjunction with a (nonprotein) coenzyme. There are various types of enzymes. Carbohydrases, for example, break down carbohydrates; transferase enzymes transfer various groups—for instance, carboxylases add or subtract carbon dioxide. The synthesis of enzymes and other proteins, and the passing on of hereditary information, is controlled by nucleic acids. There are two kinds found in the nuclei of cells: the complex DNA (deoxyribonucleic acid) and the simpler RNA (ribonucleic acid).

Secondary plant products

In addition to carbohydrates, fats, and proteins, there are several other classes of chemicals produced by plants, many of commercial value. The nitrogen-containing alkaloids, for instance, are toxic compounds used as drugs. Their function in plants is probably to deter grazing animals. Most occur in members of three families: the poppies (Papaveraceae), buttercups (Ranunculaceae), and the nightshade family (Solanaceae). Examples of alkaloids include atropine, colchicine, morphine, nicotine, opium, and strychnine.

Tannins are a group of substances frequently found in the vacuoles of cells. Most are polymers of carbohydrates and phenolic acids and have a bitter, astringent taste. Their function, like alkaloids, may be to deter herbivores. Cultivated plants containing them include tea, coffee, and bilberries.

Terpenes are aromatic chemicals that are important constituents of essential oils. Together with aldehydes, ketones, and alcohols they form the scents of plants and give rise to resins and balsams. They are often produced in response to an injury from special cells and have a protective function.

Sugar (sucrose) is a carbohydrate made and stored by plants, from which it is extracted for human consumption. In temperate regions farmers grow sugar beets, seen being harvested (below); sugar cane is the equivalent crop in tropical areas.

Belladonna (Atropa belladonna) is the source of the alkaloid atropine. Like other such compounds it is highly poisonous and is prescribed as a drug in carefully controlled doses.

Plant responses

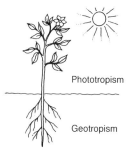

Tropisms are directional movements toward the source of a stimulus. Phototropism, the curvature toward light, and geotropism, growth toward gravity, are the two main plant responses.

All living plant cells respond directly to environmental changes (external stimuli) and indirectly to morphogenetic responses (internal stimuli). There are four principal plant responses: tropisms, which are directional growth movements in response to such stimuli as light, gravity, touch, and water; turgor movements, which are curving motions in plant organs, such as leaves or flowers, caused by changes in turgor pressure initiated by the sun's movement, day length, and touch; morphogenetic responses to day length, or photoperiodism; and the action of hormones and enzymes in growth regulation.

Experiments have shown that the responses of plants involve the production and movement within the plant of certain chemicals, the phytohormones. These compounds include auxins, gibberellins, and kinins, and are widespread in the plant kingdom. Their precise mode of operation, however, is not yet fully understood.

Tropisms

The autotrophic plant's behavior is geared to the maximum exploitation of light, which is its only source of energy. It is therefore one of the most important environmental influences on plant orientation. Motile algae, such as *Euglena* and *Volvox*, show a positive phototactic response—they move bodily toward a light source. Sessile plants, however, grow toward a light source in a movement known as phototropism. The growth curvature common to all forms of tropism occurs when there is a differential growth of cells on the opposite side of the responding organ. This is caused by the growth-regulating hormone, auxin, stimulating cell elongation on one side of the organ. In plant shoots, positive phototropic growth has been found to result from the lateral transport of auxin, which is induced by the stimulus of light. The light is perceived by a pigment called blue light receptor. Phototropism is induced most effectively by blue light but how it causes auxin transport is unknown.

Gravity also has a decisive influence on plant orientation. The main roots of plants show a positive geotropism by growing downward, whereas shoots demonstrate negative geotropism by growing away from the force. Leaves show transverse geotropism. The geotropism in roots may be controlled by abscisic acid rather than auxin (which inhibits the geotropism of root cells). Geotropic curvature in roots is not very well understood, but it is thought that gravity is perceived by metabolically inert starch grains (statoliths) in the root tips. These are displaced in the root tip tissue, for example, when the plant grows around an obstruction in the soil thus causing the statoliths to shift. They fall under gravity to the lowest side of the root where they inhibit growth and promote it on the upper side. This theory, however, does not explain why some plants which apparently lack statoliths, such as the onion (*Allium* sp.), still respond to gravity.

Chemicals also influence plant movements. Motile bacteria and algae demonstrate chemotaxis, moving bodily toward chemicals, whereas parts of higher plants show chemotropism. For example, the male gametes of some species are attracted by chemicals released by the female gametes.

Plants also respond to contact. In a move-

The day and night opening and closure of flowers (such as the night-blooming cactus *Selenicereus* sp.) and leaves are turgor-controlled. In some plants *(below right)* during the day the inner motor cells at the leaf bases become turgid and press the leaves apart. At night they collapse, and the outer cells become turgid, forcing the leaves to close.

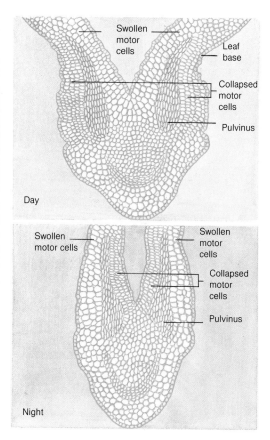

ment known as thigmotropism, tendrils and twining stems of climbing plants, such as the passionflower (*Passiflora* sp.), are induced to curl around supports that touch them. Water, or the lack of it, particularly in arid soils, also directs the orientation of root development—this is called hydrotropism.

Turgor movements

In what may appear to be a phototropic movement the leaves and flowers of some plants open and close according to the light available or follow the sun's movement through the sky. In these movements, however, light is perceived by a receptor pigment called phytochrome, which is sensitive to red light only, and the changes are due to the expansion and contraction of motor cells that gain or lose turgor pressure in response to the stimulus of the light. These motor cells are contained in an articulating structure called a pulvinus.

Plants such as the poppy *(Papaver radicatum)* track the sun's course through the day, the flower moving in a 180° curve. The closure of plants' leaves at night, or sleep movements, is called nyctinasty and is due to the action of a pulvinus in the leaf petiole. The closure of flowers at night, as in tulips (*Tulipa* spp.), is not truly nyctinastic although it is a response to decreasing light; the motor cells are not organized into obvious pulvini.

Photomorphogenetic responses

Changes in development that are not oriented toward the direction of a stimulus are morphogenetic, and those induced by light are termed photomorphogenetic. The response to day and night length is known as photoperiodism and involves the detection of light by phytochrome. Such responses include plant and seed dormancy and leaf abscission. Leaf abscission is initiated by decreasing day length in autumn. The short days alter the balance of the hormones auxin (which inhibits abscission) and ethylene (which stimulates it) and promote the development of a weakened zone at the base of the leaf petiole. Winter dormancy in plants is thought to be controlled by a balance between abscisic acid and gibberellin hormones.

The effects of phytochrome mediation—the influence of light on plant growth—also include flowering during particular day or night lengths; this ensures that all individual plants belonging to the same species flower together at an appropriate season. For example, dill *(Anethum graveolens)* needs short nights to flower, whereas cocklebur *(Xanthium pennsylvanicum)* needs long nights; in other words, dill flowers in summer but cocklebur flowers in winter.

Movements in response to touch are turgor-controlled in some plant species. The leaves of the sensitive plant *(Mimosa pudica)* are open in a normal state *(above)*. If they are touched, however, the leaves collapse *(below)*, due to sudden pressure changes in the motor cells in the leaflet bases. This response is known as thigmonasty.

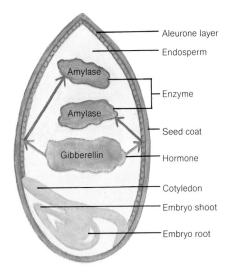

Enzyme synthesis in the seeds of grasses is induced by growth-regulatory hormones, such as gibberellin. During germination the embryonic seed releases gibberellin, which encourages the aleurone layer to produce large amounts of the enzyme amylase. This enzyme breaks down the starch stored in the endosperm of the seed, mobilizing it as a source of energy to be used by the embryo in its further growth.

Lower groups

The lower groups of organisms, that is, those which reproduce without the means of seeds, include a wide range of species, from the simplest unicellular bacterium to complex algae and fungi. In modern classification, these groups of organisms are not considered plants. Most of the groups have little more than this one reproductive feature in common. Instead of seeds they employ such diverse methods of reproduction as binary fission, conjugation, and spore formation. But despite the success and reduced risks of asexual and nonspermatophytic reproduction, most plants are spermatophytes.

The primary division of living organisms in modern classification systems is into prokaryotes and eukaryotes. Prokaryotes have relatively simple cells without true nuclei and other cell bodies, whereas eukaryotes have complex cells with nuclei and organelles, such as mitochondria and Golgi bodies. The prokaryotes consist of bacteria and blue-green algae; eukaryotes comprise all other cellular organisms. Blue-green algae are the earliest identifiable living organisms, having been found in schist that is about 3.5 billion years old.

Bacteria

Most bacteria (class Schizomycetes) are unicellular microscopic organisms. Almost all of them are heterotrophic, which means that they feed off living and decaying organisms as parasites and saprophytes, rather than autotrophic, which means that they manufacture their own food by the process of photosynthesis, as many other plants do. They occur in every possible environment and in large quantities—one drop of liquid can contain 50 million bacteria, one ounce of average soil contains more than 30 billion.

Heterotrophic bacteria include the parasitic pathogens that cause cholera, syphilis, botulism, and many other diseases. The pathogens are, however, outnumbered by the bacteria that are beneficial to us. Most nonpathogenic bacteria are saprophytes and are responsible for the decomposition of dead plants and animals, restoring essential mineral elements to the ecosystem, and preventing waste accumulation and pollution. Some live inside our bodies and assist digestion. Others are used in such industrial processes as the fermentation of alcohol to acetic acid in vinegar production and the making of yogurt and cheese.

Some saprophytic bacteria play an important role in the production of nitrogen. In well-aerated soil that is not very acid, bacteria such as *Nitrosomonas* convert ammonium to nitrite, which is itself converted by another bacterium, *Nitrobacter,* to nitrate. Nitrate is taken up by plant roots more easily than is ammonium. Nitrogen is also fixed in the root nodules of certain plants by the bacterium *Rhizobium,* which is of vital importance to agriculture. Other bacteria, such as *Pseudomonas,* reduce the nitrates in the soil to nitrogen and nitrous oxide, which are released into the atmosphere.

Autotrophic bacteria include those that photosynthesize, using light to split hydrogen sulfide rather than water. These bacteria, examples of which are *Chlorobium* and *Rhodomicrobium,* do not evolve oxygen as other photosynthetic organisms do. Other autotrophic bacteria are chemosynthetic; they are found, for example, near submarine fumaroles on the Mid-Atlantic Ridge and in areas of the Pacific Ocean Ridge near the Galapagos Islands, where they oxidize hydrogen sulfide to obtain energy.

Bacteria are usually identified and classified by cell shape, size, grouping, and flagellar arrangement. *Streptococci* are spherical and occur in groups; *Bacillus* can occur singly or in chains of rodlike cells; *Spirillum* is found singly and, as its name suggests, is spiral in shape.

Bacilli are rod-shaped bacteria, shown here *(below left)* magnified 700 times; each individual microorganism is only about 5 micrometers long. The diagram *(below right)* illustrates the main features of a generalized bacterium. Not all bacteria have flagella, but in those that do it is the whiplike lashing of this appendage that propels the organism.

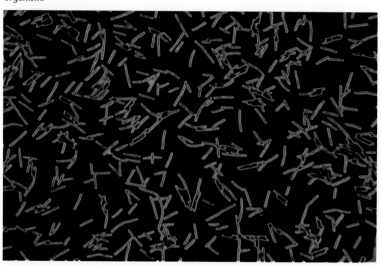

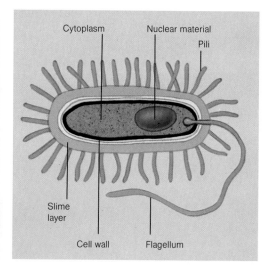

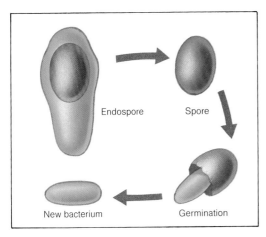

Reproduction in bacteria takes various forms. Some develop into endospores, which contain a "resting" cell called a spore. Most spores can withstand extreme changes in their environment, such as heating, freezing, or drying. They germinate when conditions again become favorable.

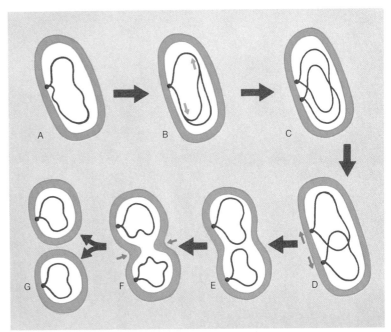

Cell division is another method of bacterial reproduction. The organism's DNA consists of a closed chromosome loop anchored to the cell wall (A). Replication of the DNA (B) results in two loops (C). The cell then elongates at the middle (D) until the loops are separated (E), and then forms a "neck" (F), which eventually divides the original cell into two new ones (G).

Bacterial cell structure

A generalized bacterial cell consists of a cell bounded by a cell wall and moved by the whiplike action of flagella. The cell wall contains polysaccharides, proteins, and lipids, but not cellulose. Gram-positive bacteria (so-called because they retain the purple color of Gram's stain) have weak walls with few amino acids. Examples are *Staphylococcus* (which causes boils) and *Streptococcus* (which, among other things, causes sore throats). They are, therefore, sensitive to certain antibiotics such as penicillin. Fortunately, most pathogens are of this type. Gram-negative forms have strong walls—they include *Salmonella* (a species of which causes typhoid) and *Escherichia* (the source of gastroenteritis).

Inside the cell, the hereditary material consists of a single loop molecule of DNA (the genophore) attached to the cell membrane, but not bounded by an envelope to form a nucleus, as in eucaryotes. Other tiny DNA fragments known as plasmids also occur in the cell and can become incorporated into the genophore when they are concerned with sexual conjugation. Plasmids can be used to transfer genes from one species to another (even from a bacterium to a higher plant or animal cell) for genetic engineering. The cytoplasm contains ribosomes for protein synthesis, and storage inclusions. These organelles are much smaller than those of eukaryotic cells. Prokaryotes also do not contain chloroplasts, as do most other plant cells.

Bacterial reproduction

Some bacteria multiply by very rapid cell fission (*Pseudomonas*, for example)—one fission can take place every 20 minutes. Others, such as *Clostridium*, produce dormant survival spores called endospores, which resist boiling and can cause food poisoning, such as botulism. Sexual reproduction in the form of conjugation is also exhibited by some bacteria. In this process, DNA passes in one direction from one bacterial cell to another via filamen-

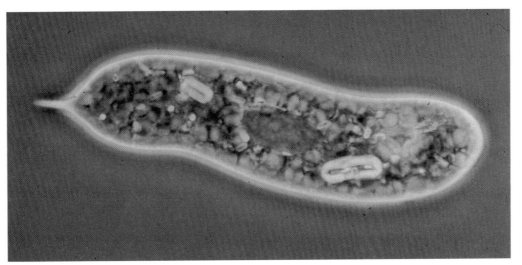

Euglena spirogyra is a photosynthetic microorganism, which propels itself using a long flagellum (at left side of the cell in this photograph). This organism, whose characteristics lie between those of plant and animal, swims toward light, which it detects with a photoreceptor located near the base of the flagellum.

Lower groups: Algae

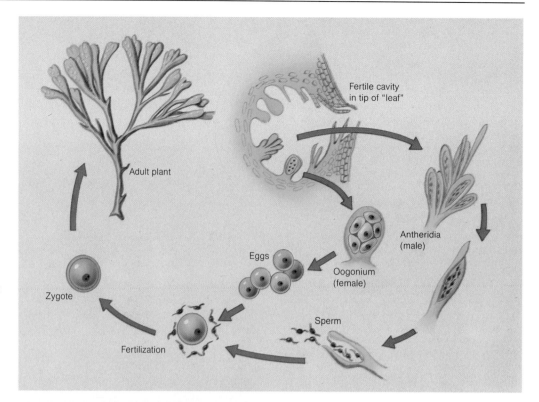

Rockweed *(Fucus)* is a common seaweed that grows on the shore between high- and low-water marks. The leathery leaflike blades *(below)* contain small cavities, in which antheridia produce male sperm and oogonia produce female eggs *(right)*. As these sex cells are released into the water, the flagellate sperm swim to the eggs and combine with them to form zygotes, from which new mature plants grow.

tous structures on the cell wall called pili. Foreign DNA can also enter a cell from the environment (transformation) or when transferred by a bacterial virus (transduction).

Cyanobacteria

The cyanobacteria, formally called blue-green algae (class Schizophyceae), are a successful and ubiquitous group of photosynthetic procaryotes. They occur in all freshwater habitats and in the sea, in the soil, as slime and gelatinous growths on rocks and manmade surfaces, as the algal partner in some lichens, and in extreme environments such as hot springs with temperatures up to 185° F. (85° C).

The cyanobacterial cell is typically prokaryotic; it differs, however, from a true bacterial cell in that photosynthetic pigments occur on the internal membranes, although they are not delimited into chloroplasts. The pigments are chlorophyll *a* (which is green), the yellow xanthophylls, blue phycocyanin, and red phycoerythrin. Different combinations of these pigments result in organisms that are blue-green, blue, black, purple, brown, red, and yellow.

Cyanobacteria usually form groups of spherical or coccoid cells, or filaments. Some filaments have complex cell types and specialized morphology, such as branching, aggregation into colonies, and three-dimensional cell division to produce simple tissues. The three main cell types are vegetative cells, akinetes, and heterocysts. Vegetative cells multiply by fission, and no sexual process is known. Akinetes are thick-walled resting spores. Heterocysts are concerned with nitrogen-fixation, a procaryotic attribute of vital ecological and economic importance; for example, the fertility of rice paddyfields depends upon the fixation of nitrogen by cyanobacteria rather than the application of nitrogenous fertilizers.

A recently discovered prokaryotic alga is *Prochloron,* which lives symbiotically inside marine tunicates (animals like the sea-squirt) on reefs. *Prochloron*'s cell structure is similar to that of a cyanobacterium, but its photosynthetic pigments include chlorophyll *b* as in green algae and higher plants. *Prochloron* could, therefore, be a key organism in the evolutionary history of land plants.

Eukaryotic algae

Algae form a heterogeneous collection of photosynthetic lower organisms ranging from single cells, multicellular colonies, and filaments to highly organized plant bodies such as seaweeds. Algae are mainly aquatic but also occur on soil, moist rocks, trees, and snow; they also occur as epiphytes and zoophytes.

In most species cellulose is contained in the cell walls. All species have no true vascular system, little tissue differentiation, and no morphological parts, such as stems, roots, or leaves. The giant seaweeds, such as giant kelp (*Macrocystis* sp.) have trumpet cells, which form tubes down the stems that transport synthesized sugar alcohols and other foods, much like the phloem system of higher plants.

The sexual reproduction of algae is unique in that the whole organism may form the gamete, or the gametes are reproduced in uni- or multicellular gametangia in which every cell becomes a gamete. The number of algal divisions varies depending on the classification system used, but in all there are at least six divisions. Among these groups, the green algae (Chlorophyta) contain motile single cells with two or four equal smooth flagella (*Chlamydomonas,* for example), colonial forms such as *Volvox,* microscopic filaments such as *Spirogyra,* and macroscopic seaweeds like *Ulva,* the sea lettuce. The pigments of this group are identical to those of mosses, ferns, and flowering plants—not surprisingly, because the land plants are believed to have evolved from the green algae. Sexual reproduction in chlorophytes varies from simple fusion of identical swimming gametes (isogamy) to advanced systems involving eggs and sperm (oogamy). Several genera have complex life cycles involving alternation of generations and are the forerunners of such systems in land plants.

The chrysomonads of the division Chrysophyta are golden-brown algae. These cells each have two unequal flagella (one with stiff hairs, the other smooth and bearing a photoreceptor), and an eyespot within a chloroplast. This arrangement is called heterokont organization. The cells are naked or are covered in siliceous scales or live inside cases (loricas) of cellulose or chitin. In many species the cells join together to form elaborate spherical or dendroid motile colonies.

In contrast, Prymnesiophytes, which are also members of the division, have two equal smooth flagella and a unique third coiling appendage (the haptonema), which may function as a chemoreceptor; the cells are covered with delicate organic scales that may become calcified to form disks and spines. Vast deposits of such flagellates in prehistoric seas have resulted in the formation of deep layers of chalk as in the White Cliffs near Dover, England, and on the French coast of the English Channel, as well as the vast Niobrara Chalk formations of Kansas.

Diatoms (Chrysophyta) are characterized by a boxlike covering composed of richly patterned silica; they are either elongate or circular, and the cells may be single or joined in chains, stars, or zigzags. Fossil deposits of diatom skeletons (diatomaceous earth) form the basis of household scouring powders and toothpaste, as well as industrial filters and insulation material. One particular type—kieselguhr—absorbs about three times its own weight of the explosive nitroglycerin to form the more easily handled explosive dynamite.

Dinoflagellates (Pyrrophyta) are brown and have a complex symmetry and an armored cuticle (periplast) of cellulose plates; some are bioluminescent and give off flashes of light when disturbed. *Gonyaulax,* for example, blooms to produce red tides; shellfish that eat these blooms accumulate a nerve toxin, which can be lethal to anyone who eats it.

Most of the other main algal divisions consist of the macroscopic thalli—the seaweeds. The red algae (Rhodophyta), which are feathery and leafy seaweeds, have no flagella or motile stages and only one chlorophyll pigment, *a.* They have, however, a predominance of the red pigment, phycoerythrin. The brown algae (Phaeophyta) comprise the bulk of the intertidal and coastal seaweeds, including the wrack *Fucus* and the giant kelps. Most of the vast array of species of brown and red seaweeds have complicated life cycles involving two or three phases.

Many of these algal species constitute the marine and freshwater phytoplankton that serve as a primary food source for zooplankton and larger filter-feeders, such as sponges, mollusks, and certain fish. Their potential and value in food and energy production is tremendous. Together the algal groups, and phytoplankton in particular, are responsible for almost half of the annual global fixation of carbon (estimated at about 23×10^9 tons of carbon per year).

Algae are not very important economically, compared with the higher plants. Certain kinds are eaten as food, especially in Wales where it is called laver bread, and in Japan, where *nori,* as it is called, is specially cultivated, dried, and eaten wrapped around rice. However, red algae provide agar and brown algae produce alginates. These jellylike substances are extracted and used in food canning; for making emulsions for use in paint, cosmetics, ice cream, medicines, and photographic film; and as a medium for growing microorganisms in the laboratory.

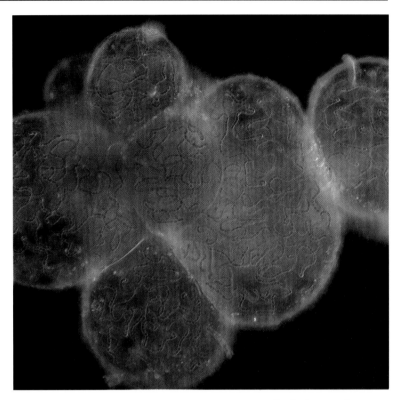

Nostoc is a cyanobacterium, here enlarged about 200 times, whose filamentous strands are held in fluid-filled sacs. It belongs to the group called Cyanophyta and, despite its name, may be black or red, as well as blue or green.

Fungi

The fungi, which make up the kingdom Mycophyta, embrace a wide variety of organisms. Nevertheless, they have several characteristics in common: they have no chlorophyll and, therefore, cannot photosynthesize, but obtain food by absorbing soluble materials as saprophytes (feeding on decaying matter) or parasites (feeding on living organisms); they reproduce by spores sexually or asexually; and they usually have cell walls that contain chitin or cellulose. Scientists now believe there are more than 100,000 species of fungi.

This broad group is divided into the division Myxomycota, which contains the slime molds (related to protozoa), and Eumycota, the "true fungi." Unlike slime molds, eumycotes are typically nonmobile and composed of branching filaments called hyphae. They include the aquatic molds and downy mildews (Phycomycota, regarded by many botanists as colorless algae) and three other groups of fungi: the so-called bread molds (Zygomycota); the powdery mildews, flask fungi, cup fungi, yeasts, and truffles (Ascomycetes); and the rusts, smuts, mushrooms, and toadstools (Basidiomycetes). In addition, there is a group of fungi—the "fungi imperfecti"—that have never been observed to have a sexual stage in their life cycle, so are difficult to classify precisely, but are usually classed as Deuteromycetes.

Myxomycota—the slime molds

These protozoalike organisms live on land in moist environments, growing on damp soil, rotting logs, and leaf mold.

The true slime molds consist of a white, yellow, or red ameboid slime called a plasmodium. An inch or so across, it moves imperceptibly, feeding on bacteria and particles of organic matter. The molds feed by engulfing their food (phagocytosis), a method which is atypical of fungi. They also differ from most other fungi by lacking a hyphal structure at any stage of their life cycle. Eventually movement ceases and the organism develops fruiting bodies known as sporangia, which produce spores. When the spores germinate, they release naked flagellated sex cells (gametes), which fuse in pairs to produce ameboid zygotes. The ameba grows into a plasmodium by feeding and nuclear division.

In cellular slime molds, the spores give rise to free-living soil amebas. When there is a large population of amebas they aggregate (but do not fuse) to form a "slug," which finally produces a sporangium. The cellular slime mold *Dictyostelium* is a common laboratory organism and has been used extensively in

Fungi, classified botanically as the kingdom Mycophyta, include a range of outwardly dissimilar organisms, from yeasts and molds to mushrooms. The major classes are shown in the diagram *(far right)*. (Remember that this system of classification may differ somewhat from systems used elsewhere.)

Resembling the antlers of a deer, a club fungus *(Xylaria hypoxylon)* grows from a rotting log in the leaf litter of a woodland floor. Mosses also take advantage of the nutrients readily available in this damp environment.

research on the physiology of ameboid movement.

Oomycota—the water molds

This group includes aquatic single-celled and filamentous forms and the downy mildews which attack some land plants. The spores or gametes are the only motile stage in the life cycles of these plants. Like the rest of the eumycotes (but unlike slime molds) they have cell walls at every phase.

Many species of those fungi known as chytrids are parasitic on algae, other fungi, aquatic angiosperms, and even fish. Among this group asexual reproduction is by uniflagellate swimming cells (zoospores), whereas sexual reproduction is by the fusion of two flagellated gametes to form a thick-walled survival spore.

Unlike most other aquatic fungi the water molds (class Oomycetes) develop an extensive branching system of hyphae, known as a mycelium. The hyphae usually contain several nuclei not separated by cell walls and are termed coenocytic. The water molds grow in water or damp soil, on seeds, dead insects, frogs, fish, and fish eggs. The downy mildews (Peronosporales) are devastating plant pathogens; *Phytophthora infestans*, for example, causes potato blight and *Pythium* attacks plant seedlings.

The oomycetes are also characterized by having cellulose only in their cell walls, asexual reproduction by biflagellated zoospores, and sexual reproduction by oogamy. This process involves the fertilization of eggs in the female sexual organs (oogonia) by male nuclei from antheridia. These and other features—typical of some algae—suggest that the oomycetes derive from algae that have lost their chlorophyll.

Zygomycota—the bread molds

These fungi are composed of the classes Zygomycetes and Trichomycetes, both of which reproduce sexually by conjugation of gametes attached to a mycelium. They form extensive molds in terrestrial habitats, occurring as bread molds, plant parasites, fungal parasites, insect parasites (the trichomycetes are parasites in the gut of arthropods only), and dung fungi; those in the order Zoopagales prey on amebae and nematodes.

The reproductive processes of the common bread molds *Rhizopus* and *Mucor* exemplify the group. In asexual reproduction, the hyphae, which grow on damp bread, produce globular sporangia that release thousands of nonmotile spores into the air. The spores land and germinate on a suitable surface to produce new, haploid mycelia. In sexual reproduction, short branches from two hyphae touch at the tips and each separates off a haploid gamete by a cross wall. The gametes fuse (conjugate) to form a diploid zygote that develops a thick protective coat, and is then known as a zygospore. After dormancy for several months the zygospore undergoes meiosis (reduction division) and germinates to produce a haploid sporangium. The spores from this structure grow into new mycelia, and thus complete the sexual cycle. The many different species of these molds are classified by the shapes and patterns of their sporangia.

Despite its attractive appearance the fly agaric (*Amanita muscaria, above*) contains a toxin that is deadly poisonous to human beings. Like the club fungus illustrated on the opposite page, a yellow slime fungus (*left*) has colonized a fallen branch as its food supply.

Ascomycetes—cup and flask fungi

The class Ascomycetes is the largest group of fungi, with more than 30,000 named species. They are mainly terrestrial and are saprophytic or parasitic. They include edible fungi, such as morels and truffles; some of the yeasts; plant parasites, such as those that cause Dutch elm disease; ergot; powdery mildews; animal parasites such as ringworm; most of the fungal components of lichens; and most marine fungi.

The unique characteristic of the Ascomycetes is the ascus—a saclike cell. This structure produces eight haploid spores (ascospores) internally by "free cell formation," which involves the spores being cut out from the substance of the cell. The ascus is formed during sexual reproduction after fusion of male and female gametes. The asci may be surrounded by a large mycelium of tightly packed hyphae that are able to form macroscopic fruiting bodies of various forms. Typical fruiting ascomycetes are the morels (*Morchella* sp.), truffles (Tuberales), and such common woodland forms as the orange-peel fungus (*Aleuria aurantia*). In these forms the asci are produced by disk-, cup-, or flask-shaped ascocarps.

Lower groups: Fungi

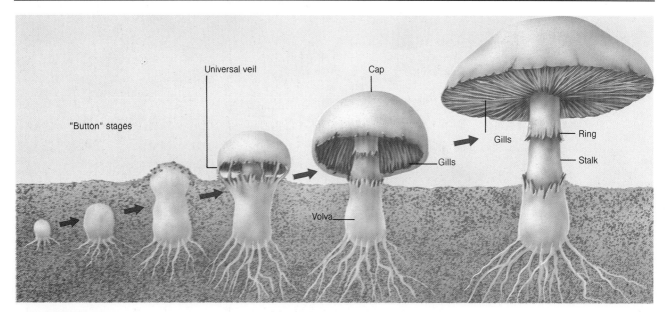

The edible field mushroom *(Agaricus campestris)* is a typical member of the class Basidiomycetes. The fruiting body—the part we eat—consists of a cap lined underneath with gills, in which the spores are formed. The development of the fruiting body, from the underground "button" stages, is illustrated above.

Aspergillus is an important Ascomycete, which has a moldlike form and reproduces asexually by single-spored sporangia (conidia). It causes considerable damage to stored grain, cloth, and other goods. It also produces aflatoxin, one of the most potent of the poisons and carcinogens.

Most yeasts are specialized Ascomycetes that grow as single oval cells. Some multiply by fission, but most undergo budding, as do baker's and brewer's yeast *(Saccharomyces cereviseae)*. In the sexual phase, two haploid yeast cells fuse to form a diploid colony in which individual cells act as asci and cleave out four or eight haploid ascospores. Yeasts can ferment sugars to produce alcohol and carbon dioxide, a property that is exploited in brewing beer, making wine, and raising bread.

Imperfect fungi

The 25,000 fungi in the class Deuteromycetes are known as imperfect fungi because they only reproduce asexually. Reproduction usually involves the production of conidia (on conidiophores) or by budding. Many of these fungi probably represent the asexual stages of ascomycetes and basidiomycetes.

The Deuteromycetes include yeastlike and mycelial forms, many of which cause diseases. A typical deuteromycete genus is *Penicillium*, which grows as a gray-green mold on rotting fruit. It reproduces asexually by the erect conidiophores producing chains of conidia. Species of this genus are well known for their commercial use in the production of the antibiotic penicillin.

Basidiomycetes

Most of the large conspicuous fungi—mushrooms, toadstools, puffballs, and brackets—are grouped in the class Basidiomycetes, of which there are more than 38,000 species. The vegetative mycelium of these fungi is an extensive, usually underground, system of septate (rather than coenocytic) hyphae. The aerial structures, such as those of toadstools, are composed of interwoven hyphae and are the spore-bearing portions of the fungus. These fruiting bodies are often brightly colored due to pigments in their cell walls.

The basidium, a club-shaped hypha specialized for reproduction, is the common characteristic of the Basidiomycetes. Single basidiospores form in four small protrusions on each basidium. Formation of the basidiospores is the end result of a type of sexual reproduction that begins when the mycelia of two different basidiomycetes merge. In each cell of this merged mycelium, called a heterokaryon, the haploid nuclei of each parent persist, sometimes for many years. Biologists think that the presence of two genetically distinct nuclei may give the heterokaryon greater ability to grow or use nutrients than either parent fungus has alone.

In the basidium, the two nuclei fuse and the

Yeasts, in the class Ascomycetes, consist of microscopic single-celled organisms. They reproduce asexually by budding, as explained in the diagram below. Yeast was probably the first plant to be domesticated, and it is still used in making bread, beer, and wine, as well as for other fermentation processes.

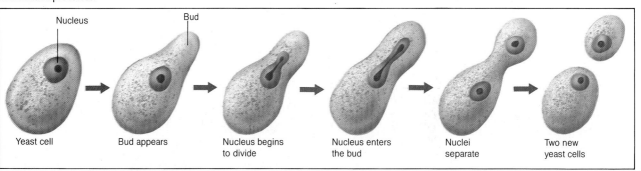

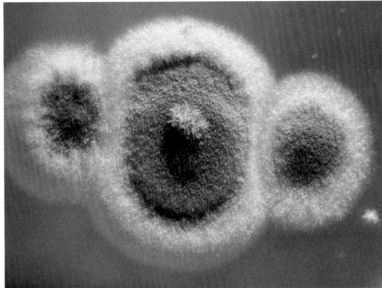

resulting zygote undergoes meiosis, forming four haploid spores. The basidia may be borne on gills, in tubes, or in fleshy masses inside the fruiting bodies as in puffballs (Lycoperdales) or outside as in stinkhorns (Phallales). A typical mushroom discharges 10 million spores per hour for several days. These are usually dispersed by wind or insects. The basidia discharge their spores by means of a little explosion that carries them a distance of 1 millimeter or so. They then drop down the space between the gills or the inside of the tubes until they reach the open air and are wafted away.

Mushrooms are often found in "fairy rings." As the underground mycelium uses the organic matter in the soil, it grows outward to find fresh food. This results in a ring of hyphae, enlarging all the time, producing fruiting bodies as it goes.

Rust (Uredinales) and smut fungi (Ustilaginales) are characterized by a variety of spore types and complex life cycles involving parasitism of several host plants. There are about 6,000 species of rusts and 1,000 species of smuts. Many live as parasites on cereals, vegetables, and flowers.

Basidiomycetes are of immense ecological and industrial importance in their mycorrhizal associations with forest trees and many other plants. Mycorrhiza is a symbiotic association between fungi and plant roots, from which both benefit. The advantages for the fungi are that the plants with which they are associated pass on photosynthetic products to them, and in turn the fungi provide the plants with mineral nutrients, which they can extract more efficiently from the soil. Many pioneer plants of poor soils are extremely dependent on these mycorrhizal fungi. Plants may form mycorrhizal associations with many fungal species; birch, for example, has a large number of basidiomycetes with which it will grow.

Antibiotics such as penicillin are produced naturally by molds, although many can now also be synthesized in the chemical laboratory. The photograph *(above left)* shows three colonies of *Penicillium chrysogenum* cultured on a nutrient jelly.

When a single raindrop falls on top of an earth star fungus *(Geastrum triplex, above right)* it explosively ejects a cloud containing many thousands of spores that are dispersed by the wind, before germinating to produce new plants.

A bracket fungus is named after its horizontal fruiting bodies, which stick out from the bark of a tree. The yellow, or sulfur, bracket fungus *Laetipurus sulphurus (below)* forms a cascade of brackets that grow up to 1 foot (30 centimeters) across.

Mosses and liverworts

The mosses and liverworts are a group of simple green plants found throughout the world—they are among the most primitive of land plants. Classified botanically as bryophytes, the species number more than 9,000 for mosses and about 8,000 for liverworts. They require water to complete their life cycles; for this reason most of them frequent damp, shady habitats. Some bryophytes (such as the moss *Polytrichum*) can, however, withstand periods of drought.

Plant structure

Most bryophytes form cushions or layers of vegetation no more than 6 inches (15 centimeters) high for mosses and less than .5 inch (1.4 centimeters) high for liverworts. But the largest moss, the Australian species *Dawsonia*, grows up to 3 feet (91 centimeters) tall. All are attached to the ground or to a substratum, such as tree bark, by rootlike threads called rhizoids. In liverworts rhizoids generally comprise one cell each, whereas in mosses they are made up of several cells.

Nearly all bryophyte structures that grow above ground contain chlorophyll, as well as other pigments, and are green; the chlorophyll is located in disk-shaped chloroplasts clustered in certain cells. The moss plant consists of a system of shoots along which leaves are arranged alternately. In most moss species the leaves have a characteristic central midrib, but otherwise vary enormously in shape and size, from the pointed, tooth-edged leaves of *Orthotrichum* to the broad leaves of *Mnium*, which have unbroken edges and long hairlike extensions at their tips.

Liverworts are more diverse in appearance than mosses and are described as either thalloid or leafy. A thalloid liverwort consists of a flat, platelike structure (the thallus) with rhizoids on its underside. The thallus may have a scalloped edge, as in *Lunularia,* or may be deeply divided into lobes, as in *Conocephalum.* In leafy liverworts, which comprise more than 80 per cent of the total number of species, the thallus is divided into leaflike structures arranged up the "stem" in ranks of two or three. Most liverwort leaves have no midrib.

Alternation of generations

Bryophytes are of considerable botanical interest because they are the most primitive plants to display alternation of generations. According to this system, the plant's life cycle takes place in two stages or generations: the gametophyte generation, which is responsible for producing sex cells, and the sporophyte generation, whose function is to make and

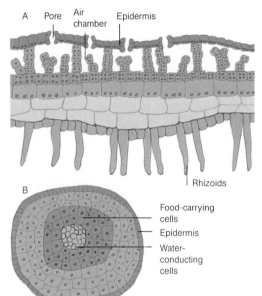

A liverwort thallus (A) has no leaves, stem, or roots. It is simply arranged into upper layers of chloroplast-containing cells and lower, empty ones. Fine rhizoids attach it to the soil. A moss stem (B) has a primitive vascular arrangement of non-nucleated, water-conducting cells, surrounded by nucleated ones which carry organic compounds.

The spore-filled capsules at the tip of the stalks of mosses *(below)* form the sporophyte generation.

Nonflowering plants: Mosses and liverworts

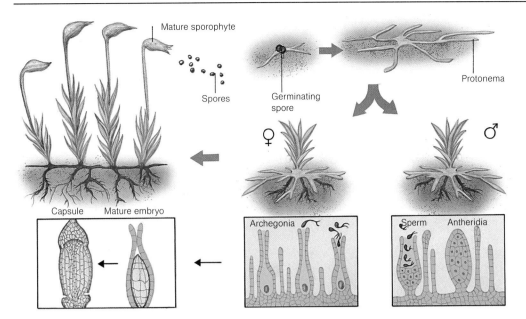

The capsules on a mature moss plant release spores. A spore develops into a threaded mesh (protonema), which, from a bud, gives rise to a male or female gametophyte plant. Club-shaped sperm are released from the antheridia of the male plant and swim to the archegonia on the female plant where they fertilize the egg. Each egg develops into an embryo that eventually becomes the spore-producing capsule.

disperse large numbers of asexual spores. In flowering plants the gametophyte generation is confined to specific parts of the flower, but in bryophytes it takes the form of the obvious main green "body" of the plant.

Anatomically, bryophyte sex organs vary little between species. Male sex cells (sperm) are made in antheridia, and female sex cells (called eggs) in archegonia. The antheridia are thin-walled sacs, which, when mature, contain sperm, each of which bears a pair of long whiplike flagella. The fully developed archegonium containing the egg is usually flask-shaped with a long neck.

For sexual reproduction to take place, a sperm must swim to an egg in an archegonium and fuse with it. Water is thus essential to the process of fertilization. The journey of the male cells is facilitated by a "spreading agent," which when released from the antheridium, lowers the surface tension of the water and thus aids the flow of the sperm. The fertilized egg develops into a sporophyte, which is incapable of an independent existence. Instead it grows as a semiparasite on the gametophyte, although in some bryophytes, such as the unusual *Anthoceros,* it contains chlorophyll and so is capable of photosynthesis. The typical sporophyte of a moss or liverwort consists of a flattened foot, a long stalk or seta, and a capsule containing thousands of spores. The methods of spore release from the capsule vary, but most depend on the drying of the capsule and an inbuilt spore-ejector mechanism.

Spores that fall on a suitable environment germinate and develop into a new gametophyte generation. As well as reproducing sexually, this gametophyte also has the capacity for asexual reproduction—that is, leaves or other parts broken off the plant may develop into new, independent gametophytes.

Evolution and ecology

The evolution of bryophytes is not well understood. Botanists are still not certain whether

bryophytes evolved from a kind of alga or whether they represent a retrograde evolution from pteridophytes—the ferns and their allies.

Ecologically, bryophytes are important because they can survive in inhospitable conditions and because they are often among the pioneer species that colonize land that has been laid waste by catastrophes such as fire or earthquake. Once established, bryophyte colonies inhibit soil erosion and promote the retention of soil moisture. In swamps, the *Sphagnum* moss plays a leading role, building organic soil and eventually making it sufficiently firm and nutrient-rich to support shrub and tree species.

Because many bryophytes have specific nutrient requirements, it is possible to tell whether habitats are acid or alkaline, or contain high levels of nitrogen or phosphorus, by the types of bryophytes that inhabit them. Similarly, bryophyte species can, by their presence or absence, indicate whether there is a high level of sulfur dioxide in the atmosphere.

Cup-shaped structures found commonly on liverworts contain special bodies called gemmae. These structures are organs of vegetative reproduction. The gemmae are distributed to new habitats by splashes of water into the cups, or by attaching to the feet or fur of passing animals. In a suitable environment they develop into a new gametophyte plant.

Club mosses and horsetails

Club mosses and horsetails, together with quillworts, ferns, and whisk-ferns, are known as pteridophytes (fern plants). These primitive plants were much more abundant at least 300 million years ago in the Pennsylvanian period than they are today. Their ancestors reached an enormous height—more than 100 feet (30 meters)—and were dominant over large areas of land. Superficially, many of them resemble mosses, and they are sporophytic—that is, they reproduce from spores instead of seeds—but they differ from mosses in that they have a vascular system.

Club mosses and quillworts

The members of the class Lycopsida are grouped into three living orders: Lycopodiales (with about 200 species) and Selaginellales (more than 700 species) both contain the club mosses, and Isoetales (about 70 species) contains the quillworts. Most lycopsids are found in the tropics and subtropics but some occur in temperate, desert, arctic, and alpine regions. They usually grow on the ground, although a few species live on other plants commensally (when they are known as epiphytes). The quillworts are found worldwide, and most live largely below the ground level, with only the tips of their sporophylls (fertile leaves) showing.

Most club mosses creep or trail although some have erect stems. The roots generally grow directly from the stem, and some species produce rhizophores, which grow down from the stem to the soil with true roots issuing from their tips. The leaves, which photosynthesize, are small with an unbranched midrib—those of *Lycopodium* are needlelike. In most species the leaves are spirally arranged, but in some species of *Selaginella* they are attached in four rows—two on the upper side and two laterally. The leaves may all be the same size or there may be a regular pattern of large and small leaves. In *Selaginella,* a membranous scale (ligule) grows at the base of each leaf, the function of which is not known. Ligules are also found in quillworts during the development of the sporophylls.

In addition to the ordinary green leaves, club mosses have fertile leaves (sporophylls). In contrast to the club mosses all the leaves of the quillworts are sporophylls. These leaves, which project above the soil level, are the only photosynthetic part of the plant; the rest survives as a corm (a condensed stem) underground.

In most species of club mosses the sporophylls are grouped at the top of the stem to form a cone (strobilus). Each sporophyll has a single large spore receptacle, or sporangium, on its upper surface near the base of the leaf.

Stag's horn club moss *(Lycopodium clavatum)* is named after its long white stalks, which usually have two fertile cones (strobili). Its trailing stems, which are often 10 feet (3 meters) long, give rise to fertile and sterile branches. The stalks grow from the fertile branches. Their spore leaves (sporophylls) are smaller than those of the sterile and fertile branches and grow more closely against the stem. This club moss is homosporous—its spores are all identical.

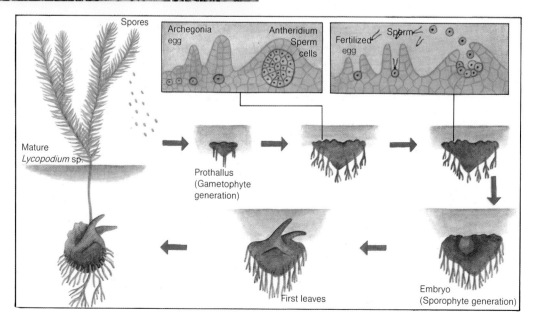

Alternation of generations in some club mosses involves the underground germination of a spore. The resulting prothallus (the gametophyte) has female sex organs (archegonia) and male ones (antheridia) on its upper surface. When the sex cells (gametes) mature, the necks of the archegonia open and the antheridia release the sperm cells whose walls rupture. The sperm "swim" to the archegonia where they fertilize the egg. The developing embryo begins the sporophyte phase.

The sporangium may be globular or kidney-shaped. *Lycopodium* bears sporangia that are identical and produce one type of spore only. It is, therefore, described as homosporous. *Selaginella* and the quillworts, however, are heterosporous, having both small microsporangia and larger megasporangia.

In club mosses the spores are released when they are ripe—the sporangium splits open and the spores are dispersed by the wind. In quillworts the spores have to wait until the sporangia decay before they can be released. In suitable conditions, each spore germinates in the ground to form a tiny prothallus. In some species, the prothallus lives underground in a symbiotic association with mycorrhizal fungi from which it gets its nutrients. In other species the prothalli are surface-dwelling and photosynthetic. In *Selaginella* and the quillworts, the prothalli are endosporic—that is, they develop within the spores.

The life cycle of club mosses consists of two alternating phases—the gametophyte generation and the sporophyte generation. The gametophyte phase begins with spores that germinate into prothalli and ends with the fertilization of the female gametes by the male sperm.

The sporophyte generation begins with the fertilized egg (zygote), which develops into the spore-producing plant. In *Lycopodium* the male and female gametes are produced in the same prothalli. *Selaginella* and quillworts, however, have minute microprothalli, which have a single male organ (antheridium) and much larger megaprothalli, which are produced by the megaspores and which have several female organs (archegonia). Sperm and eggs are thus produced on separate prothalli. To achieve fertilization the sperm must swim through a film of water from the microprothallus to the archegonia on the megaprothallus.

Reproduction among the club mosses is, however, not always sexual—in some species of *Lycopodium,* leafy stem structures called bulbils detach themselves from the plant and develop into new plants.

Horsetails

The horsetails (class Sphenopsida) are today represented by a single genus—*Equisetum*—with about 25 species. They are found worldwide, except in Australia and New Zealand, and in a varied range of habitats. These plants have long underground rhizomes that give rise to aerial stems that are usually 4 to 24 inches (10 to 61 centimeters) in height. The stems normally contain chlorophyll, but some species alternate the growth of special fertile stems without chlorophyll with green sterile ones.

The stems are grooved and have thick silicified walls. They are simple (unbranched) and carry whorls of slender branches, which are simple or have further whorls of branches. The small leaves, too, are produced in whorls, each cluster being fused into a tubular sheath, except for the tips, which form a serrated edge around the margin of the sheath. Most of the leaves have no chlorophyll, and photosynthesis is carried out by the green stems.

The reproductive organs of horsetails are grouped together to form a terminal cone, or strobilus. Each strobilus consists of a central axis bearing whorls of mushroom-shaped (peltate) sporangiophores, each with several sporangia on the lower surface. The sporangia split to release the spores. Each spore has four long strips called elaters, which coil and uncoil in response to humidity changes and assist its dispersal in water and wind. Some of the spores germinate to form male prothalli, which bear antheridia, whereas others produce female prothalli with archegonia. In others still, antheridia and archegonia may be borne successively on the same prothallus. Fertilization is achieved in the same way as in the club mosses; the horsetails also develop by an alternation of generations.

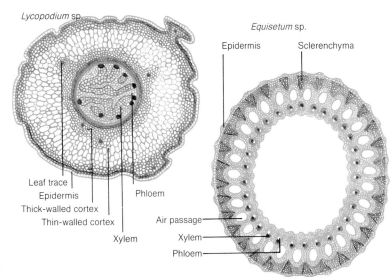

A cross section through the stems of club mosses and horsetails reveals their primitive vascular systems. *Lycopodium* has a central cylinder which contains lobes of xylem interspersed with phloem. *Equisetum* has a hollow stem ringed by small xylem and phloem bundles.

The great horsetail *(Equisetum telmateia)* produces pale photosynthetic fertile stems with strobili at their tips in early spring. Later in the year sterile stems with green leaves develop, which are seen here. This plant grows in damp woods and on hedge banks.

A young fern, the sporophyte generation of its reproductive cycle, develops from a prothallus, the gametophyte generation. The new fern is nourished by the photosynthesizing prothallus, but when it can photosynthesize for itself, the prothallus dies.

Ferns

The ferns (class Filicopsida) first appeared in the Devonian period, more than 350 million years ago, and have flourished ever since, particularly in forests. They are a large group of plants, containing approximately 10,000 species, which now inhabit a variety of environments. Most are ground-dwelling, preferring damp, shady habitats, but some are epiphytic, growing attached to the stems or leaves of other plants; others can survive in more exposed situations, and a few have an entirely aquatic life style.

Ferns vary widely also in their appearance, ranging from the giant, treelike ferns, which can reach a height of 65 feet (20 meters), to tiny, mosslike species that grow on rocks. They all belong to the division Pterophyta and are, therefore, closely related to club mosses, horsetails, and quillworts. Like them, ferns have alternating sporophyte (asexual) and gametophyte (sexual) generations. The class is roughly divided into the "primitive" eusporangiates, such as adder's tongue (*Ophioglossom vulgatum*) and moonwort (*Botrychium lunaria*), and the "higher" leptosporangiates (which contain the most common ferns, such as maidenhair fern, *Adiantum* sp.); the subclass between, Osmundidae, shares features of both groups. The class Osmunda includes the complex-leaved royal fern, *Osmunda regalis*. (Remember that there are several classification systems; the system used here may differ slightly from others.)

Structure

In most ferns the stem is usually a short, thick stock, as in the common male fern *(Dryopteris filix-mas)*, or a long rhizome, as in bracken *(Pteridium aquilinum)* and epiphytic ferns. The stock grows almost completely buried in the ground, with an ascending growing point above ground surrounded by a close spiral of leaves, resembling a whorl, and often forming a basketlike tuft. Roots usually grow from the backs of the leaf-bases and on the lower surface of rhizomes, which grow horizontally above or below the ground, bearing single leaves at intervals along the upper surface. The growing point of rhizomes is surrounded not by leaves, but by scales or hairs, which are usually brown and sometimes glandular.

The leaves (fronds) of most ferns are spirally arranged and known as megaphylls. When they are in bud, in most species, they are rolled up like a shepherd's crook, or crozier. Like the stems, they are normally covered (at least when young) with scales or hairs. They contain chlorophyll and photosynthesize. Small adjustable pores called stomata, through which gas exchange takes place, are located in the epidermis of the leaf, and water and nutrients in solution are carried through the leaves and stems by xylem and phloem tissue.

Spore production

Some ferns are able to reproduce vegetatively (asexually) by means of creeping rhizomes or by bulbils produced on the leaves, but propagation is usually by spores formed in sporangia. In most species, the sporangia are located on the lower surface of ordinary leaves; others bear them on the axis of the leaves or at their tips. Still others have separate fertile leaves, such as the royal fern.

Usually, many sporangia are grouped together to form a sorus, many sori occurring

The sporangia on most species of ferns are borne on the underside of the fronds, grouped into sori. On buckler fern, each sorus is partly covered by a kidney-shaped indusium. On lady fern the indusium forms a flap, whereas on holly fern it is reduced, forming a dimpled disk which barely covers the sorus. The sori of common polypody have no indusium. The sporangia of maidenhair fern develop on the margins of the leaves and are protected by the leaf curling over them. Those of tree ferns are contained in cuplike receptacles sometimes covered by an indusium which splits to uncover the sporangia.

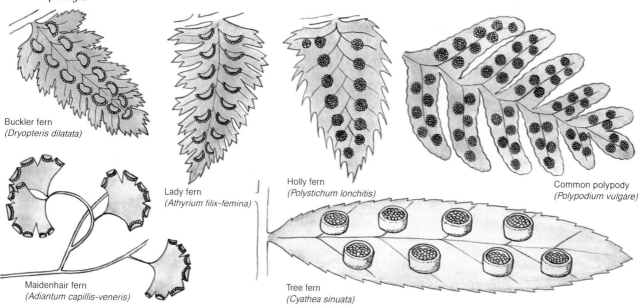

Buckler fern
(*Dryopteris dilatata*)

Lady fern
(*Athyrium filix-femina*)

Holly fern
(*Polystichum lonchitis*)

Common polypody
(*Polypodium vulgare*)

Maidenhair fern
(*Adiantum capillis-veneris*)

Tree fern
(*Cyathea sinuata*)

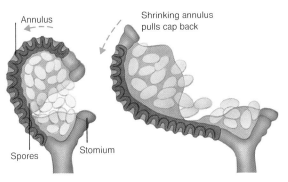
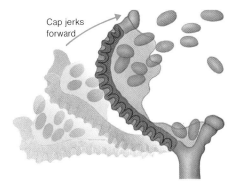

The sporangium of a fern splits at a point called the stomium when the spores are ripe. The surrounding ring of cells (the annulus) shrinks, pulling the cap back. When the tension reaches a maximum point, the cap jerks back to its original position, shooting the spores into the air.

on each leaf or leaf-segment. Each sorus is normally situated on a cushion-shaped structure, called the receptacle, directly above a vein that supplies it with nourishment. In some species, such as common polypody (*Polypodium vulgare*), the sori are not protected, but in most ferns each sorus has a cover (the indusium), which is most commonly linear, kidney-shaped (reniform), or mushroom-shaped (peltate).

The sporangium is usually stalked and has a head that is often shaped like a biconvex lens, with a ring or annulus of thickened cells almost completely surrounding the margin. Thin-walled cells occupy the remainder of the margin, flanking a weak point called the stomium, which eventually ruptures to release the spores. The sporangia contain a large number of spores; *Christensenia*, for example, releases about 7,000 spores from each sporangium, and adder's tongue more than 15,000. These spores contain chlorophyll.

Sexual reproduction

If a spore lands in favorable conditions it germinates a few days after it is released, forming a tiny green prothallium (the gametophyte). Hairlike roots (rhizoids), which form on the lower surface of the prothallium, supply the plant with water and nutrients, and in some species the prothallia photosynthesize. Most prothallia, however, are also mycorrhizal.

On the underside of this cushion of cells are the female reproductive organs (archegonia), each one consisting of an egg cell and a narrow, cylindrical neck that projects from the surface of the prothallium and is filled with neck canal cells. The male reproductive organs (antheridia) are normally located on the same prothallium, toward its base. Inside the spherical wall of each antheridium is a cavity containing sperm cells, each producing a single sperm with several long whiplike hairs (cilia or flagella) with which it propels itself through water.

Because the only way the sperm can reach the archegonia is by swimming, fertilization takes place only when the prothallium is covered with a film of water. This moisture causes the antheridia to split open, releasing the sperm cells whose walls dissolve, setting the sperm free. Simultaneously, the canal cells of the archegonia disintegrate, releasing a chemical that attracts the sperm and leaves the neck open. The sperm travel up the neck canal where one enters the egg and fuses with the nucleus.

Following fertilization, the resulting zygote begins its development attached to the prothallium and is nourished by it. But when it has formed its first root stem and leaf and is able to photosynthesize and thus manufacture food itself, the prothallium dies.

Most ferns are termed homosporous, which means that their spores are all alike. But the water ferns (those belonging to the orders Marsileales and Salviniales) are termed heterosporous, which means that they produce two different types of spores from different sporangia—micro- and megasporangia—on the same plant. In these ferns, the tiny microspores germinate to form the male prothallia, which bear antheridia, and the larger megaspores give rise to female prothallia, with archegonia.

Tree ferns (families Cyatheaceae and Dicksoniaceae) are found in mountain forests in tropical areas such as the South Pacific, Malaysia, and parts of Australasia. They can reach 65 feet (20 meters) in height and usually have unbranched stems that end in a crown of leaves. They first appeared in the Jurassic period about 170 million years ago.

Seed ferns and cycads

The seed ferns (order Pteridospermales) and the cycads (Cycadales) are both very ancient groups. The seed ferns are known only from fossils dating from the mid-Devonian to early Cretaceous periods, 360 to 130 million years ago, when they had a worldwide distribution. About seven families can be identified, but the number of species is not known due to the fragmentary nature of the fossil record. The cycads, however, have living representatives, although they date back to the late Triassic period, 180 million years ago. They comprise 9 genera and about 75 species, all of which are now confined to the New and Old World tropics, whereas once they too were widespread.

Together the seed ferns and cycads make up the class Cycadopsida and are grouped as gymnosperms along with the other cone-bearing plants. They both have manoxylic wood—it is soft and spongy with wide parenchyma rays. Their leaves are large and frondlike and tend to branch pinnately (like a feather). In addition, their seeds are radially symmetrical (the same on all sides).

The seed ferns

The fossils of seed ferns are relatively common although incomplete. They reveal, nonetheless, that the leaves of these plants were large and fernlike (and consequently became broken at some stage during fossilization). It has also been found that the cambium in the stems formed large amounts of secondary xylem, unlike the cycads.

The plant parts have all been found but have not been joined together and consequently have been given separate scientific names. This has to a certain extent obscured the relationship between the various organs and made it even more difficult to build up a complete picture of the plants. This problem is compounded by the existence of a large number of different species. Enough has been discovered, however, to suggest that the appearance of seed ferns ranged from *Lyginopteris*, a scrambling plant with long thin stems and equally branched fertile fronds (sporophylls), to *Sphenopteris*, a large upright plant that resembled a tree fern.

Seed ferns evidently had neither cones nor flowers but produced seeds that were naked and not enclosed in an ovary. The presence of seeds places these plants among the gymnosperms and sets them apart from the ferns that they resemble so closely in appearance. The seeds were contained in leafy cupules on the ends of leaves or on special nonleafy branches (much like those on tree ferns). The seeds occurred either singly or in numbers up to about 70 in one cupule. These containers were shaped rather like a modern tulip flower, with the seeds on stalks inside them. It is thought that as each seed matured, the stalk elongated and carried the seed upwards to the mouth of the cupule, where it could more easily be dispersed.

Pollen sacs were also borne on the fronds. Without evidence, however, the method of pollination can only be guessed at, but it has been suggested that the female gametes were wind-pollinated. Nothing has been learned from any fossils of details of fertilization or embryo development.

Cycads

From what can be seen of their structure it is thought that the seed ferns formed an intermediate evolutionary stage between the true ferns and the conifers. Despite the fact that cycads show many primitive features, which suggest a close relationship with the seed ferns, they are considered to be intermediate between the seed ferns and the flowering plants, and some botanists believe that the flowering plants may have originated from the early cycads.

Most species of cycad are similar in appearance. The stem, or trunk, resembles that of a palm and can vary in height between species, up to about 60 feet (18 meters), as in the Australasian cycad *Macrozamia hopei*. Most, however, are only about 3 feet (91 centimeters) high and some species even have an underground stem, or tuber, such as *Stangeria* spp., from South Africa. The stem develops little secondary xylem but contains large amounts of parenchyma mixed with other conducting cells. It is usually unbranched and scarred with the bases of fallen leaves. The leaves, which grow in a crown at the top of the trunk, resemble fern or palm fronds and may vary in length from 10 feet (3 meters) in *Cycas* sp. to 2 inches (5 centimeters), as in *Zamia pygmaea*.

Male and female flowers occur on separate plants (they are dioecious) in the form of a cone at the top of the plant, up to about 3 feet (91 centimeters) long and 10 inches (25 centimeters) thick. The cones are formed in a spiral

The male cones of the cycad *Encephalartos frederici-guilelmii*, of southern Africa, are formed from tight spirals of microsporophylls.

from fertile leaves (sporophylls). In the male flower the sporophylls are known as microsporophylls because pollen sacs containing the microspores, or pollen, grow on their underside. The microsporophylls may be up to 20 inches (50 centimeters) in length.

The sporophylls of the female cone are known as megasporophylls and are much smaller than the microsporophylls, being only 6 to 8 inches (15 to 20 centimeters) long. They are also conelike, except in *Cycas,* when they form a loose aggregation of sporophylls. The ovules, which may grow to the size of a hen's egg, or larger, are formed in numbers from two to eight on the lower half of the megasporophyll (as in *Cycas*) or may hang from under the sporophylls. Pollen released from the microsporangia is transferred to the ovules by the wind or by insects. Up to six months may pass before fertilization occurs, and the seed may take a year to mature.

Because the flowers are terminal, growth in male plants continues from an axillary bud at the base of the cone. The female flowers of most species of cycads are similar except in the case of *Cycas,* in which the apical meristem is unaffected by the flower and it continues its normal growth through the middle of the flower.

Cycads have adapted to survival in a dry climate by being xeromorphic—the long-lived ones live up to 1,000 years. The leaves have a thickened epidermis with sunken stomata. Apart from a long taproot, the roots are short and grow as coralloid masses (groups of short, thickened interwined roots), which contain symbiotic cyanobacteria, such as *Anabaena.*

The cycads have little economic value, although several species are grown as ornamental garden or house plants.

The cycad *Encephalartos latifrondia* indicates how palmlike these primitive plants are, except for their small stature—3 feet (91 centimeters) high at the most.

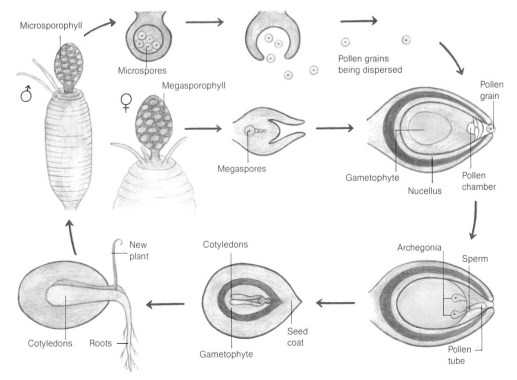

Pollination of cycads occurs when microspores, which have developed into pollen grains, are carried by the wind to the female megasporophyll. This body contains a gametophyte with two archegonia. The pollen grows a tube that fertilizes a gamete in one of the archegonia (this may take six months). The developing seed may take a year to germinate and is nourished by the starchy foodstore of the gametophyte.

Cone-bearing plants

The cone-bearing plants comprise the biggest group of nonflowering plants—the division Pinophyta (the gymnosperms). This group bridges the gap between the ferns and the angiosperms—the flowering plants. The structure of gymnosperm cones and the details of their reproduction may indicate how angiosperm flowers evolved; both groups may have had a common ancestor at some stage in their evolution—gymnosperms are known to date from the Permian period, about 275 million years ago.

Gymnosperm characteristics

The name gymnosperm applies to the seed (*gymnos* is the Greek word for naked), which is not enclosed within an ovary. This means that the extensions of the ovary wall found in angiosperms—the style and stigma—are absent. In most gymnosperms the fertile leaves (sporophylls) grow in tightly grouped formations called strobili, which in many species are conelike.

The gymnosperms can be divided into three classes, based on the anatomy of their wood, the shape of their leaves, and the structure of their seeds: the Cycadopsida, the Gnetopsida, and the Coniferopsida.

The Cycadopsida comprises the seedferns and cycads, together with other fossil orders. These fern- and palmlike plants are described in the previous article.

The class Gnetopsida contains three odd groups—the welwitschias, the ephedras, and the genus *Gnetum*. Despite being gymnosperms these plants have close affinities to the angiosperms, particularly in their wood structure.

The Coniferopsida consists of the ginkgos, the conifers, and the yews. These plants have secondary wood that is relatively dense with small parenchyma rays. The wood is simpler than angiosperm wood and contains the elongated cells called tracheids but does not in most gymnosperms have true vessels. Pith rays found in the phloem and xylem tissues transport food from the leaves to the trunk for storage and carry water away from the trunk. Some conifers, such as spruce (*Picea* sp.), have no fibers.

The wood parenchyma of gymnosperms is usually associated with resin. These cells may form a network from the resin canals, as in pines (*Pinus* spp.), or a column of cells that fill with resin. The leaves are usually needle- or scalelike, but may be fan- or paddle-shaped. The leaves or veins branch dichotomously—that is, they branch equally into two and then two again, and so on.

The odd group

The plants in the class Gnetopsida differ widely in appearance but all share some features—for example, they are all woody and contain conducting vessels. Also, the microsporangia have a perianth surrounding them, as may the female megasporangia. The megasporangia themselves are arranged in compound strobili, or inflorescences. No fossils (apart from pollen grains) of any of the three living genera have been found, so relationships between them cannot be determined.

The presence of vessels in the secondary xylem may be a characteristic shared with angiosperms but does not indicate a close relationship with them—it is, rather, an example of convergent evolution. They have evolved from pitted tracheids, whereas angiosperm vessels are derived from tracheids with ladderlike (scalariform) thickening.

Found only in the deserts of southwestern Africa is the unique welwitschia (*Welwitschia mirabilis*), the only species of the order Welwitschiales. This peculiar plant has a large,

The welwitschia *(Welwitschia mirabilis)* of the Namib Desert appears at first sight to be an unlikely relation of the cone-bearing plants. But the dark patches on top of the plant, centrally placed between its two leaves, are groups of cones. These cones are similar to those of conifers in that those of the male plants bear pollen grains and those of the female plants carry an ovule. Pollination is mainly by means of insects.

Nonflowering plants: Cone-bearing plants

The pretty, fan-shaped, deciduous leaves of the ginkgo *(Ginkgo biloba)* stand in sharp contrast to the needlelike evergreen leaves of other members of the order Coniferopsida. This tall, spreading tree *(above)* has ovules at the end of short shoots on the female trees and catkins of male sporophylls on the male trees. The photograph above shows young male catkins.

globe-shaped, underground stem, with two strap-shaped leaves that can be 2 to 3 feet (61 to 91 centimeters) wide and often twice as long. The leaves continue to grow throughout the life of the plant (more than 100 years), and are continually split into ribbons by the hot, desert winds. Male and female strobili grow on different plants in the form of a cone covered by scales or bracts. The female cones are larger than the male cones and are scarlet when mature. Pollination is mainly by insects.

The order Ephedrales comprises the small, thin xerophytic shrubs, which are found in North and South America and from the Mediterranean eastward to China. The plants have small, scalelike leaves and are dioecious (with male and female organs on different plants). The male strobili are solitary and form a conelike inflorescence. The female strobili grow in groups of two or three and are wind-pollinated. Depending on the species, the seeds are either dry and winged or brightly colored and fleshy; they are wind- or animal-dispersed, respectively.

The species of the order *Gnetales* are usually found in tropical rain forests. Most are woody climbers (lianas) although some species grow as shrubs or trees. The leaves are remarkable in that they look just like dicotyledon leaves with a central midrib, net venation, and a broad blade (lamina). The cones resemble small upright catkins and are generally dioecious. The females are wind-pollinated and produce seeds in which the embryo is covered with a protective stony layer and surrounded by a fleshy outer layer. The seeds are dispersed by birds.

The ginkgo

The ginkgo *(Ginkgo biloba)*, or maidenhair tree, has distinctive two-lobed, fan-shaped leaves. The leaves are identical to those that fell from ginkgos millions of years ago, probably in the Triassic period, when some were preserved in mud and fossilized. Then it had a worldwide distribution, whereas now the ginkgo is found naturally only in Asia. It remains the sole representative of the order Ginkgoales.

The mature tree has a broad spreading crown that reaches a height of about 80 feet (24 meters), with deciduous leaves. Male and female reproductive organs are found on separate trees. The male microstrobili, which bear pendulous microsporangia, are like catkins produced at the end of short shoots. The pollen is transferred by the wind to the ovules, which are also produced at the ends of short shoots. The sperm released from the pollen is large with a spiral band of flagella—a primitive feature. Unlike all other members of the class Coniferopsida, the sperm is motile and swims to the female gamete. The mature ovule falls from the tree in autumn and fertilization may take place while it is on the ground. The seed develops a stony inner layer and an outer fleshy one, which gives off a smell like rancid butter. This smell may be attractive to the animals that are potential dispersers of the seeds.

Conifers

The true cone-bearers belong to the order Coniferales. Most living species are trees, and only a few can properly be described as shrubs. Some of the tallest of living trees are included in the order—the Californian redwood *(Sequoia sempervirens)*, for example.

The order contains six families, the largest one being the pines (Pinaceae). The others are the cypresses (Cupressaceae), the plum yews (Cephalotaxaceae), the redwoods and swamp cypresses (Taxodiaceae), the araucarias (Araucariaceae), and the podocarps (Podocarpaceae). Most are found in the Northern Hemisphere, although the last two families occur predominantly in the Southern Hemisphere.

The wood of conifer trunks has tracheids with large pits in their walls, but no vessels or pores. Resin canals are common, found sometimes in the wood but mostly in the leaves and cortex. The branches are generally regularly arranged up the trunk to give most conifers their distinctive pyramidal appearance.

Most conifers are evergreen and in some species, notably the monkey puzzle tree *(Araucaria araucana)*, the leaves may remain on the tree for up to 15 years. A few species drop their leaves every autumn—larch *(Larix* sp.), for example. The leaves are usually needle-

A cross section through a female cone from a conifer *(right)* reveals the cone scales, or megasporophylls, each containing one ovule. In these ovules one megaspore develops into a gametophyte. In the male cones *(far right)* the microsporophylls each contain several microsporangia, which develop two or three microspores; these eventually become pollen grains.

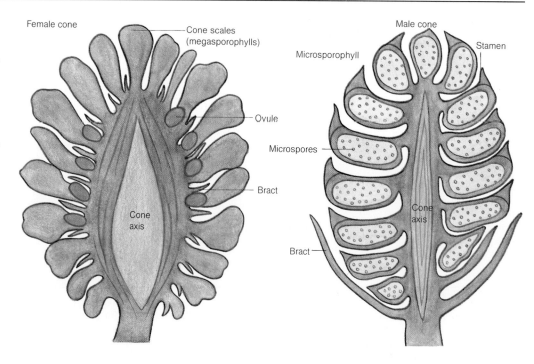

shaped, scalelike, and small, but may be broad, as in podocarps (*Podocarpus* spp.), whose leaves are 12 inches (30 centimeters) long and 2 inches (5 centimeters) wide.

Conifers are adapted to growing in boreal zones and on mountains, where there may be a lack of water when the ground is frozen in winter, although there are exceptions—such as the swamp cypress (*Taxodium* sp.) of the southern United States. In addition, their pyramidal shape ensures that snow is more likely to slide off the branches than collect on them, possibly breaking them. The conifer leaves are constructed to resist drying out, with a thick, waxy cuticle and, usually, a small size. The needle shape is most resistant to frost. These xeromorphic characteristics equally help those conifers that live in arid places such as sand dunes, as do many species of pine (*Pinus* sp.). The stone pine *(Pinus pinea),* for example, is a familiar sight on bare stony Mediterranean hills where water is in short supply.

Conifers can be monoecious or dioecious. In all species, however, the cones are unisexual. The female cones, or megastrobili, contain two ovules. In some species of podocarp, however, there is only one ovule in the cone. The cones usually consist of a central axis with large woody bract scales attached to them, each scale carrying a seed. But not all cones take this form; in some species, such as juniper, the bract scales are so reduced that the cones resemble berries (which are used for flavoring gin), whereas in the podocarps the cones look like small plums.

The male cones, or microstrobili, tend to be smaller than the female ones and simpler. In the larch the male cones turn from bright red to dark red to chestnut-brown when ripe. They contain microsporophylls with pollen sacs—each cone may produce up to 10 million pollen grains. Pollination is by means of the wind in all conifers. The pollen grains that land on the ovule develop a pollen tube with one male

Pines (*Pinus* spp.), along with most other conifers, are well adapted to living in cold environments. The pyramidal shape of these trees means that heavy snow will slip off the branches, preventing them from breaking.

nucleus, which fertilizes the female gamete.

The roots of conifers are inhabited by mycorrhizal fungi. The fungal hyphae grow either between the cells of the cortex (when the association is known as ectotrophic mycorrhiza) or inside the cells (endotrophic mycorrhiza). In the podocarps the fungi grow in special root nodules. These fungi supply the trees with minerals, especially nitrogen.

The yews and associated species

The evergreen shrubs and small trees that make up the order Taxales grow only in the Northern Hemisphere, except for the monospecific genus *Austrotaxus,* which is confined to the island of New Caledonia in the South Pacific. There are five living genera, the sixth (extinct) genus *Palaeotaxus* having provided the basis upon which the order was separated from the true conifers (its seeds were not present in cones, but attached to part of a branch). The yews belong to the genus *Taxus,* and other genera include the torreyas (*Torreya* spp.), *Amentotaxus,* and *Pseudotaxus.*

The plants vary in size from a height of 100 feet (30 meters) with a diameter of only 10 feet (3 meters) as in *Torreya* to a height of 65 feet (20 meters) and a diameter of 22 feet (7 meters) as in the yews. The massive trunk of yews may not be a single trunk but represents several fused smaller ones.

Yew leaves are flattened and shaped rather like a Roman sword. They are usually about 1 inch (2.5 centimeters) in length, but reach about 3 inches (8 centimeters) in the Californian nutmeg *(Torreya californica).* The leaves contain one vascular bundle only, which has special transfusion tissue on each side of it that facilitates the movement of materials between the leaf and the bundle. The leaves spread apart in two rows along the stem. The branches grow out horizontally from the main trunk and form a rather dense umbrella under which little will grow.

Unlike the wood produced by conifers, the wood of these plants does not contain resin canals or wood parenchyma cells. The yew leaves do not have resin canals, but in the other genera resin sacs occur in the leaves and flowers. The tracheids of the secondary wood have abundant spiral thickening and may account for the elasticity of the wood, which made it so popular for bows.

Whereas the ovules of most of the other gymnosperms, except *Podocarpus,* are grouped, the ovules of these plants are solitary, growing at the end of a small branch and not in a cone. They are surrounded by stony integuments and a succulent cup, or aril. The red fleshy aril is eaten by birds, which consequently disperse the seeds. The aril is the only part of the plant that does not contain the poisonous alkaloid taxin.

Yews are dioecious. Pollen is produced by microstrobili that take the form of cones or scales. It is transmitted to the female flowers by the wind, and the seeds are produced a few months later. The seeds have two cotyledons that persist after germination for about three years; they are similar in shape to the true leaves, but somewhat larger.

Redwoods (*Sequoia* spp.) are among the largest trees in the world, generally reaching a height of about 200 to 275 feet (61 to 84 meters) and a diameter of 8 to 12 feet (2.4 to 3.6 meters).

Cypresses (*Cupressus* spp.) are a familiar sight among the olive groves of the Mediterranean countries and are one of the few examples of conifers that do not grow in cold, mountainous regions.

The bright red cups, or arils, of yew trees (*Taxus* spp.) surround the ovules. The color attracts birds that eat the aril and later disperse the seed.

Flowering plants

The development of the flowering plants is possibly the greatest success story in the evolution of the earth. From the time of their appearance in the middle of the Cretaceous period, about 130 million years ago, they have flourished and diversified so that today they dominate nearly all the terrestrial plant communities, from tropical forests to temperate deciduous woodlands, and from grasslands to deserts. Even the sea has flowering plants of some kind. They are also the prime producers of food for the animals of the earth (but not the sea), not least for humans who rely on them as food for themselves and domesticated livestock.

Flowering plants are grouped into the Angiospermae, or Magnoliophyta, about 250,000 species of which have been named and described. These plants are further arranged in 2 groups—the dicotyledons, or Magnoliopsida, with about 190,000 species, and the monocotyledons, or Liliopsida, with about 60,000 species. (There are, however, several systems of plant classification; the system described in this book may differ slightly from classification systems used elsewhere. For examples, *see The World Book Encyclopedia* articles PLANT [A classification of the plant kingdom] and CLASSIFICATION, SCIENTIFIC.) In addition to the thousands of "natural" species, millions of forms, varieties, and cultivars have been artificially created by plant breeding programs, in order to decorate gardens and homes, and to increase the abundance and the extent of crops.

The appearance of angiosperms

The first fossils with rudimentary angiosperm characteristics, such as pollen grains, date from the Triassic period, which began about 245 million years ago. The first true angiosperms, which possibly evolved from an ancestral cycad, were most likely to have been tropical trees with flowers—possibly similar to those of a present-day magnolia—and with large, fleshy edible fruits that were attractive to birds and other animals.

These appeared in the early part of the Cretaceous period, about 130 million years ago. During this period—the climax of the Age of Dinosaurs—the land flora underwent a great change, and many of the new plants would be recognizable today. As well as the magnolias, there were willows, oaks, poplars, and sycamores. The ubiquitous grasses that make up such a familiar part of the modern landscape did not develop until Cenozoic times, when grasslands, and their accompanying running mammals, spread at the expense of the forests.

But the exact way in which the angiosperms evolved and the steps that occurred between the appearance of the first flowering plants and the production of the diversity that exists today remains a mystery. The answer lies in the huge gaps in the fossil record—possibly due to the fact that flowering plants evolved in

The flowering parts of plants vary, from the primitive plants in which they are complex and many, to the advanced plants in which they are simple and reduced. Sunflowers *(Helianthus annuus),* like all members of the Compositae family, are higher dicotyledonous plants with only their central disk florets capable of forming seeds. The yellow, outer, petallike florets serve to attract pollinating insects and birds.

parts of the world that were subsequently submerged—and until these gaps are filled, the precursors of the angiosperm plants will remain unknown.

The characters of success

The whole design and life of the angiosperm flower, whether it is a showy bloom or an inconspicuous grass flower, is devoted to the production of viable seed from which the next generation can grow. Flowering plants differ from the other major class of plants—Gymnospermae—by having their female sex cells (ovules) enclosed in an ovary; the gymnosperms have naked seeds. In addition, the seeds that develop from angiosperm ovules are enclosed in a fruit. But the most important feature of the angiosperm life style is that the generation of gametophytes, on which the insurance of genetic diversity depends, is reduced to the activities of a few cells only, and confined to the safe depths of the flower.

The first flowering plants were hermaphroditic—that is, they contained male pollen-producing anthers and female ovule-containing ovaries in the same flower. This arrangement is still found in most modern angiosperms, but it seems likely that the primitive plants possessed genes that induced a state of self-sterility—the pollen from one flower was not capable of fertilizing the ovules of the same flower—which made a transfer of pollen between the flowers of different individuals of the same species essential.

Animals were originally responsible for this necessary process of the transfer of pollen from flower to flower—a role they have now played for millions of years. Only in specialized advanced states of angiosperm development has animal help been replaced by obligatory self-pollination, seed production without the need for fertilization, or by wind-pollinating mechanisms. Insects were the first pollinators to aid plants in their need—they were tempted by the pollen produced by the plants, which is rich in nutrients. In time they turned to the sugary nectar and, as the insects entered the flower in search of the nectar, they brushed against the rough-surfaced pollen, which stuck to them. On the next visit to a flower of the same species, the pollen was transferred to the receptive stigma and carried to the ovule, where fertilization was effected.

As the plants' systems evolved, petals became brightly colored and patterned to attract birds and guide insects to the nectar. Many of these patterns are visible only in the ultraviolet wavelength, a part of the light spectrum in which insect vision can operate. Flowers also developed scents that were attractive to insects and birds. The opening of flowers and scent production became synchronized to the active periods of pollinators, and as the reliance of specific pollinators increased, so the structure of the flower became more complex. Those flowers that do not rely on animal associations for pollination do not have color, nectar, or scent because they do not need to attract animals.

Wind, animals, the explosive mechanisms of the seeds, and occasionally water, are the prime dispersers of angiosperm seed. Because the seeds have a self-contained food supply, they are also freed from the necessity of an external food source when they start to germinate; they can also remain dormant for long periods if they need to. Nevertheless, angiosperm seeds develop extremely quickly in comparison to, say, gymnosperms. A dandelion, for example, takes six weeks from seed germination to seed dispersal from a mature plant. A conifer, on the other hand, can take up to two or three years. These characteristics have made angiosperms highly adaptable and able to diversify, and account for their enormous success on this planet.

Frangipani (*Plumeria* sp.) is a simple tropical flower. It attracts pollinating insects with its bright pink or white and yellow flowers, and also with its distinctive scent.

Tiger lilies (*Lilium tigrinum*) are monocotyledonous flowers, and, like many others, have exposed flowering parts. The six pendulous, lobed structures are the stamens (male). There are three stigmas (female) and three compartments in the ovary. The large, showy flowers are pollinated by insects that push past the protruding stamens to reach the nectar, picking up pollen as they move, and transferring it to the stigmas.

The flower is the organ of sexual reproduction in plants. The most obvious parts are the petals, which form the corolla, and sepals, which form the calyx; together these two constitute the perianth, mounted on the receptacle. The female pistil is made up of the stigma on its style above the ovule-containing ovary. The male stamens consist of pollen-bearing anthers on supporting filaments.

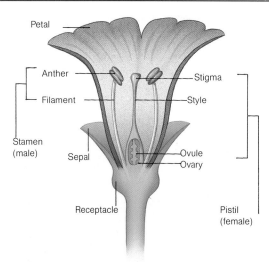

Sexual reproduction

If plants increased by vegetative means only—through budding, runners, and so on—the processes of evolution and adaptation could not take place, because each new plant would have exactly the same genetic makeup as its parent. The element of change relies on sexual reproduction, in which each new individual inherits genetic material from both of its parents. Sexual reproduction has the additional advantage of enabling the individual embryos to be carried away from the parents and to grow some distance away, perhaps in a more favorable environment. This system helps to prevent the plants from becoming overcrowded and assists the spread of the species.

Flowers

The part of a plant specialized for sexual reproduction is the flower. Here the male and female gametes—pollen (male) and ovules (female)—are produced and come together to produce seeds, each containing an embryo plant. In some species, such as buttercup (*Ranunculus* spp.), all the flowers are hermaphrodite, having both male organs (stamens) and female organs (carpels). Other plants have unisexual flowers, producing either male or female gametes, but not both. Separate male and female flowers may both be borne on the same plant (monoecious), as in hazel (*Corylus americana*), or male flowers may be produced on some plants and female flowers on others (dioecious), as in willows (*Salix* spp.). Species with male, female, and hermaphrodite flowers, either on the same plant or on different plants, are known as polygamous.

Flowers may be borne singly or may be grouped together to form an inflorescence. Solitary flowers are seen in anemone (*Anemone patens*), whereas the flowers of Virginia cowslip (*Mertensia virginica*) form an inflorescence. In some species, such as oxeye daisy (*Chrysanthemum leucanthemum*), what appear to be single flowers are in fact inflorescences made up of many tiny individual flowers closely grouped together.

Carpels

The top of a flower stem is often enlarged and forms a platform (the receptacle) upon which the other floral organs are located, arranged in whorls, spirals, or both. In the center are the carpels, collectively forming the pistil. Each carpel consists of a sticky stigma at the top of a stalk (the style), which joins a hollow ovary containing one or more ovules. Buttercups have a single ovule in each carpel; some orchid carpels contain half a million ovules each.

Species with only one carpel in each flower are rare; they include the members of the pea family. Sometimes the carpels are separate as in buttercup, but in most species two or more carpels are fused together, having a common ovary. The edges of the fused carpels may project into the ovary but not reach the center, producing a unilocular ovary; however, more frequently, the carpel edges meet at the center of the ovary, dividing it into as many loculi (chambers) as there are carpels. The styles and stigmas may also be entirely or partly fused together, or they may remain separate.

Each ovule consists of an ovoid mass of cells called the nucellus, which is attached to the ovary wall by a short stalk (the funicle). Protective layers (integuments) cover the entire nucellus except for a small opening called the micropyle. Inside the nucellus is the embryo sac, usually containing eight nuclei. Three are situated near the micropyle, the middle one of these forming the ovum (egg); three are at the opposite end to the micropyle; and two are in the center of the embryo sac. These two central nuclei usually fuse together to form the

Male and female structures show up well in the flower of a hybrid fuchsia. The female stigma protrudes beyond a cluster of five stamens, some of which bear powdery pollen.

endosperm nucleus or secondary nucleus.

Stamens

Surrounding the carpels are the stamens. They may be separate, or all or some may be joined. Their number is constant in some species, but variable in others. Each stamen has a swollen head (the anther) and usually a stalk (the filament), which carries nutrients from the plant to the anther. The central part of the anther is called the connective, and attached to this are pollen sacs containing pollen mother cells, each of which divides twice (by meiosis) to form a tetrad of four pollen grains. In some species the grains are dispersed in these groups of four, but usually they separate within the pollen sacs. When ripe, the anther opens, generally by means of two longitudinal slits, and the pollen sacs are ruptured, releasing the pollen grains.

The shape of pollen grains varies, but often they are round or oblong. The outer wall has distinctive pores and may also possess spines, ridges, or other features. These characteristics are remarkably constant in any individual species so that it is usually possible to identify a plant by examining its pollen grains. Each grain has two nuclei, one of which forms the generative cell, and the other is the tube nucleus. The generative cell usually divides to form two male gametes or sperm cells; therefore the mature pollen grain contains three nuclei, one vegetative and two generative (gametes).

Petals and sepals

The stamens are usually surrounded by a corolla, formed of petals. These are not directly involved in the reproductive process, but can be of great assistance to the plant because large, brightly colored petals attract insects that may carry out pollination, transferring pollen grains from stamens to stigmas. In some species the petals are joined to form a corolla tube or trumpet shape.

A calyx, composed of sepals, normally forms the outermost part of the flower, protecting the other organs, especially before the bud opens. The sepals may be joined into a calyx tube. Both petals and sepals are of constant number in some species, and variable in others. Usually the sepals are green, but some flowers, such as those of bluebell and crocus, have sepals that are petaloid—they are just like the petals in size, shape, and color, and the two whorls are known collectively as the perianth. Sometimes, as in gladiolus *(Gladiolus tristis)*, the parts that appear to be petals are in fact large, showy sepals. Some plants, particularly those pollinated by the wind, have neither corolla nor calyx, because they do not need to attract insects. These "naked flowers" grow in plain, tassel-like inflorescences, called catkins, and can be found on plants such as alders, poplars, and willows.

Before the male and female gametes can unite the pollen grains must be carried from the stamens to the carpels (pollination). Pollen may be deposited on the stigma of the same flower or on the stigma of a different flower on the same plant. Both these processes are forms of self-pollination. When the pollen is transported to the stigma of a flower on a different plant, cross-pollination occurs. Self-pollination perpetuates the characteristics of the parent plant, whereas cross-pollination introduces new genetic combinations, bringing the possibility of better, stronger plants.

Self-pollination

In plants where self-pollination normally occurs, the stamens are usually located so that the anthers are above the stigma, and stamens and stigma ripen simultaneously. The ripe pollen often simply falls onto the receptive stigma below, although sometimes the process is more complex, involving movement of the

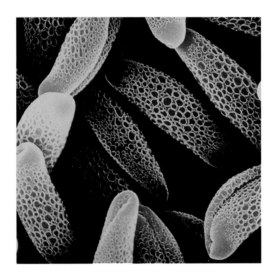

Pollen grains vary in texture and shape—depending on the plant species—from smooth and spherical to pitted and slipper-shaped *(left)*. For fertilization to take place, pollen from the male anther has to get to the female stigma. If this takes place in the same flower (A) or between two flowers on the same plant (B), self-pollination occurs. Transfer of pollen to a flower on another plant (C) leads to cross-pollination.

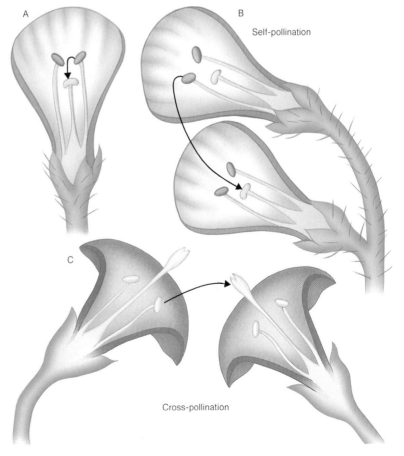

Flowering plants: Sexual reproduction

Two different designs of flowers of the same species is termed heterostyly. The "pin" form *(left)* has a long stigma and anthers deep in the flower's corolla, whereas in the "thrum" form *(right)* the positions of the organs are reversed. When an insect visits either flower to feed, pollen left on it by the anthers of one type is transferred to the stigma of the other type.

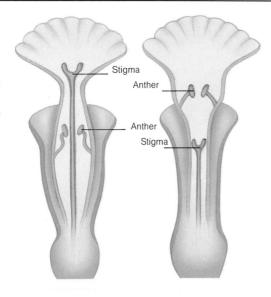

stamens or stigma.

To prevent self-pollination taking place, many species have stamens and stigmas that do not ripen at the same time. Self-pollination is impossible in dioecious plants, because male and female flowers are borne on separate plants. Sometimes pollen may fall onto a ripe stigma of the same flower but then fail to bring about fertilization, because its genes are incompatible with those of the ovule. Such plants are described as self-sterile.

Another device discouraging self-pollination is called heterostyly, which can be seen in the primrose *(Primula vulgaris)*. The flowers of this species are usually of two types (dimor-phic): a "pin-eyed" form, with the stigma at the mouth of the corolla tube on a long style and the stamens halfway up the corolla tube, and a "thrum-eyed" form, with the stigma halfway up the tube on a short style and the stamens located at the top of the corolla tube. The pollen from one form sticks to an insect visitor and is then deposited on a stigma of the other form, as the stamens of one type are at the same level as the stigma of the other. The pollen grains of the two types are different and the plants are self-sterile, a fact which suggests that heterostyly is, by itself, a rather inadequate method of discouraging self-pollination.

Cross-pollination

Cross-pollination is brought about by the wind, by insects (or birds, bats, or other animals), or, in a few aquatic plants, by water. Wind pollination, the simplest method, is also the most wasteful: vast amounts of pollen are produced in an attempt to ensure that some of it lands on a ripe stigma. The male flowers are often borne on dangling catkins, where they are shaken by the slightest breeze. This form of pollination is particularly common among trees, grasses, rushes, and sedges. The flowers are usually small and inconspicuous, with large anthers exposed to the wind on long, slender filaments, and exposed feathery stigmas to catch airborne pollen. The flowers produce neither nectar nor scent.

Insect pollination is less wasteful, but often much more complex. The flowers attract pollinators by means of large, brightly colored petals, and often by a strong scent. This is not always a sweet perfume; some flowers pro-

The main kinds of fruits can be divided among simple, aggregate, and multiple types. Simple fruits may be fleshy, as in drupes, pomes, and berries, or dry, as in nuts and the "seeds" (actually fruits) of grasses and some trees. An aggregate fruit, like blackberry, has separated seeds in a pulpy flesh (carpel), all derived from one flower. Multiple fruits, like pineapple, derive from several flowers that condense during development to form a single seed-bearing structure. Figs are another example of multiple fruits.

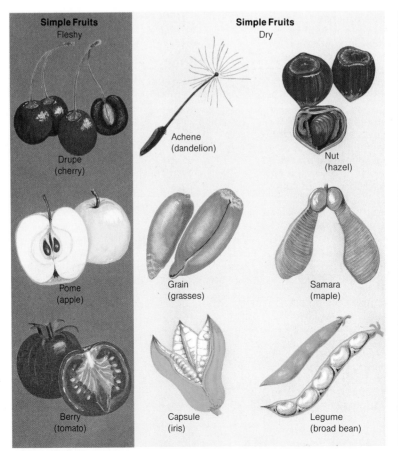

duce a stench like that of rotting flesh or dung and, thus, attract carrion-feeding or dung-feeding insects. A supply of nectar frequently serves as an invitation to insects, or the visitors may come for the pollen itself. The petals often have bright spots or lines pointing the way to the flower's nectar-store; such markings are known as nectar-guides. Nectar located deep within the flower can only be reached by insects with a long proboscis, such as bees, butterflies, and moths; flies and beetles, which have short tongues, can take nectar from flowers only where it is easily accessible. Thus different flowers are visited by different types of insects. Pollen becomes attached to the hairy legs and body of a feeding insect and is then brushed onto the stigma of the next flower visited.

Some more complex pollination systems include mechanical traps that actually hold the pollinating animal long enough for the pollen to be transferred. The jack-in-the-pulpit *(Arisaema triphyllum)* attracts flies into an enclosed chamber by the smell of carrion. The flies can escape only when the structures at the mouth of the chamber wither, by which time the flies are thoroughly dusted by pollen during their struggles. A similar device makes sure that the pollen is shaken onto the female structures on another plant. Some orchids, such as the early purple orchid *(Orchis mascula)*, have spring-loaded anthers that snap shut on a visiting bee. The bee's struggle to escape dislodges the pollen sack onto its back. It then flies to another flower of the same species and the pollen is released on the stigma. This is an excellent example of coevolution.

Water pollination is rare, even among plants that grow in water. Most pollen becomes ineffective when wet, and even in aquatic species that are pollinated by water, it is often the male flowers, and not the individual pollen grains, that are carried by the current to the female flowers, the pollen thus remaining dry.

Fertilization

When a pollen grain of the same species lands on a ripe stigma it swells as it absorbs water, sugar, and other materials from the stigma. The correct pollen then germinates or grows a tube down the style toward the ovary. The pollen tube, containing two male gametes, grows through the canal and enters an ovule by way of the micropyle. The tip of the pollen tube penetrates the embryo sac and releases the two male gametes. One enters the ovum and fuses with its nucleus, forming a zygote from which an embryo plant develops. The other usually fuses with the central nuclei of the embryo-sac, producing nutritive tissue called endosperm, which nourishes the embryo. Fertilization is thus achieved.

A hoverfly feeds on nectar and has grains of yellow pollen on the hairs of its head and body. By this means, the pollen will be transferred to another flower of the same species and will fertilize it.

Oranges ripen on a tree in Australia. The fruit, which develops from the ovary of a fertilized flower, contains several seeds, although horticulturalists have also created seedless varieties of oranges and other fruits.

Fruit and seeds

Following fertilization, many changes take place in the carpel. Each ovule becomes a seed, its integuments forming the protective seed coat (testa), and the ovary becomes a fruit, its three layers of wall forming the pericarp. Sepals, petals, and stamens usually wither, although their dry remains often persist, sometimes attached to the fruit.

There are three main kinds of fruits, classified as simple (that is, single), aggregate (clusters), and multiple, with various types within these categories as illustrated on the opposite page.

Seed dispersal

Plant seeds are the product of sexual reproduction, and their dispersal determines where succeeding generations will grow. Seed dispersal is the process that enables a species to expand its territory and helps to reduce the competition for resources between the parent plants and the seedlings.

The seed is often contained in a unit of dispersal, such as a fruit. The embryo inside the seed lies dormant while dispersal takes place, a factor that is as important to the embryo's survival as the protective structures of the seed or fruit. In many cases, the seed, fruit, partial, or whole plant is structurally adapted for dispersal by animals, wind, or water, or for self-dispersal.

Animal dispersal

The dispersal of plant seeds by insects, birds, and mammals may be carried out in various ways, all equally haphazard. The fruits or seeds may be eaten for their nutritional content and the undigested seeds may subsequently be voided elsewhere. Many fruits and seeds have some edible part, whether it is the fleshy outer part of a seed, as in the peony (*Paeonia* sp.), or more commonly of a fleshy fruit, such as a tomato.

Birds feed largely on fleshy fruits retained on the plant. They are attracted primarily by color (they are particularly sensitive to red) because they have little or no sense of smell. The different colorations of immature fruit, reinforced by an unpleasant taste, warn the birds of unripe fruit. This allows the young seed embryo more time to develop. The red currant (*Ribes rubrum*) is an example of this type, as is the mistletoe (*Phoradendron flavescens*), the seeds of which stick to birds' bills and are later scraped off.

The fruit of wild mangoes (*Mangifera* spp.) are eaten by the biggest fruit-eating bats, and the trees are adapted for dispersal by them. The mangoes hang away from the dense foliage and often have a strong rancid smell. These adaptations are necessary because the bats are less maneuverable than birds, are nocturnal and color-blind, and have relatively unsophisticated sonar systems for their species. They do, however, have teeth and a good sense of smell. They suck the fruit dry, spitting out the hard seeds.

Fruit enclosing seeds dispersed primarily by terrestrial mammals are often shed and are easily eaten on the ground. An example is the spiny durian (*Durio zibethinus*), which also has an exceptionally powerful smell. Its oily fruit is eaten not only by humans but also by rhinoceroses, orangutans, elephants, and even carnivorous cats. Its seeds, however, are toxic, which encourages the animals to excrete them as quickly as possible.

Some birds and some of the smaller mammals collect and store nuts for consumption at a later stage. Squirrels and rats store beechnuts and hazelnuts underground, and jays cache acorns in tree hollows. Where hoards are forgotten, seeds are left to grow.

Ants disperse seeds of some *Datura, Euphorbia, Cyclamen,* and *Primula* species and are drawn to them by attractant oils. Small structures (elaiosomes) impregnated with these oil food-substances are appended to the seeds and are easily detached. Alternatively, the oils may be generally distributed in the outer layers of the seeds, which are borne close to the ground. The ants eat the oil substance and leave the rest.

Another way in which seeds are dispersed is when seeds and fruits become attached to animals. They may be transported in pieces of mud that cling to the animal, as are the seeds of sweet grass (*Glyceria* sp.), which stick to the feet of passing waterbirds. Some plants produce a sticky mucilage that helps them to adhere to passing animals. Mistletoe berries eaten by birds also stick to their beaks and plumage, eventually falling off or being removed when the bird preens itself some distance away from the parent plant. Other plants produce fruits and seeds that bear hooks by which they become attached to animals. Stickseed (*Hachelia* sp.) has a small nutlike fruit with barbed prickles that catch in fur. Burdock (*Arctium* sp.) has hooked bracts below the flower, which remain on the fruit.

Wind, water, and self-dispersal

Wind is another important way in which seeds are scattered; it operates best on the outer edges of plant communities and in fairly open environments. The dustlike seeds of orchids can weigh as little as 0.001mg and are released in quantities of hundreds of thousands; a single fruit capsule of the *Cynorchis* orchid may release 4 million seeds of similar weight. Some of the heavier fruits and seeds have specific adaptations to increase their surface area relative to their weight and assist in flight. Such adaptations include the "parachute" of the dandelion (*Taraxacum officinale*)—actually a ring of fine hairs, called a pappus, attached to the seed—the downy fruit and style of virgin's-bower (*Clematis* sp.), and the hairy tufts of poplar (*Populus* sp.) and willow (*Salix*

The structure of seeds determines the way in which they will be dispersed. Winged seeds, such as those of the maple, are light and can be carried by the wind. So too can clematis and milkweed seeds, aided by their light hairy tufts, which act as a parachute mechanism. Other seeds are contained in edible fruits and rely on animal consumption for them to be released and dispersed. The barbs and spikes on some seeds also depend on animals for their dissemination, becoming attached to fur or clothing. Some fruits, such as the squirting cucumber, blow their seeds out when they release internal turgor pressure.

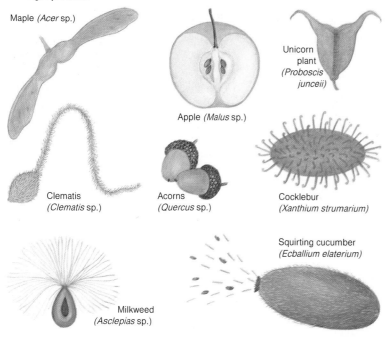

Maple (*Acer* sp.)

Apple (*Malus* sp.)

Unicorn plant (*Proboscis junceii*)

Clematis (*Clematis* sp.)

Acorns (*Quercus* sp.)

Cocklebur (*Xanthium strumarium*)

Squirting cucumber (*Ecballium elaterium*)

Milkweed (*Asclepias* sp.)

Flowering plants: Seed dispersal

sp.) seeds. The fruits of ash *(Fraxinus excelsior)* and maple *(Acer* sp.) have wings that slow their fall and help to widen their distribution. Tumbleweeds *(Salsola kali)* are more radical seed dispersers. The whole plant is blown about, scattering its seeds as it rolls. The poppy *(Papaver* sp.) ripens its fruit on long stalks; when these dry out they are caught by the wind, and the seeds are catapulted out through holes in the fruit.

Some plants rely on water to disperse their seeds. The coco-de-mer *(Lodoicea maldivica)* produces seeds that can weigh up to 45 pounds (20 kilograms) and that float with the help of trapped air. Water-repellent oils and chemicals also enhance buoyancy. But water can also assist a plant's self-dispersal of seeds. The impact of a raindrop on the calyx of the sage, *Salvia lyrata*, flexes the delicate flower stalk, which, on regaining its undisturbed position, throws off the attached nutlets.

Specially weakened tissues in plants can be ruptured by high local water pressures, throwing the seeds in all directions. For example, the touch-me-not *(Impatiens biflora)* has pods that burst open at the slightest touch, scattering seeds everywhere. Excessive water absorption can also create torsions sufficient to release seeds, as in *Vicia* and *Lupinus* species. Torsions in the awns of storkbills *(Erodium* spp.) provide enough force for the attached seeds to bore themselves down into the ground. Certain plants have explosive fruits that, when ripe, rupture and release their seeds in a high-pressure jet of fruit pulp, as does the squirting cucumber *(Ecballium elaterium)*.

Some sedges rely on the wind to disperse their seeds. When airborne, the feathery seedcases can travel for many miles.

The bright colors of fleshy fruits attract birds to them. Their seeds have a hard casing that may need the animal's digestive juices to act on them and change the structure of the seed coat so that germination is induced.

Coconuts are the seed carriers of the tropical seashore palm tree *Cocos nucifera*. These fruits can weigh several pounds but are dispersed by the sea, floating on the water by means of the air trapped inside the shell. The seed's germination may be delayed for up to two years, allowing enough time for the coconut to reach land.

Vegetative propagation

Most higher plants reproduce by seeds or spores or develop a new plant from part of themselves asexually by vegetative propagation. There are only a few annual plants that reproduce vegetatively, but most herbaceous perennials do reproduce in this way, as well as relying on seeds for propagation.

Most plants that reproduce vegetatively are outstandingly successful in terms of numbers and areas colonized.

Plantlets growing from the edge of *Bryophyllum* sp. (right) drop to the ground and root to form new plants. In the same way, the purple bulbils developing in the papery bracts of crow garlic or *Allium vineale* (below) eventually fall to the ground, where they start growing. By these means the plants can multiply without relying on seed dispersal.

Advantages and disadvantages

Vegetative reproduction involves no exchange of genetic material and consequently every new plant is identical to the parent. The lack of potential variability of the offspring, and the absence of the dispersal mechanisms that operate in seed production, are the two major disadvantages of vegetative propagation.

On the other hand, asexual reproduction uses less energy than sexual, and in a harsh environment this factor may be of vital importance. Whereas seeds must carry their own food supply to provide the developing seedling with sufficient energy to become established, the vegetatively developing plant has abundant resources at its disposal from the parent plant. The new plant is thereby not as vulnerable to the problems of competition as a seedling is.

A seed is also vulnerable during its dispersal and may land in an area unsuitable for its growth—this rarely occurs in vegetative propagation. Seeds and seedlings are further disadvantaged in that any damage to them results in death or, at least, a malformed plant. But the vegetatively produced plantlet can withstand damage more easily because it has adequate supplies from the parent for repair.

Colonization of a wide area by vegetative means is rather slow because these plants tend to form clumps—such as buffalo grass *(Buchloë dactyloides)*. But competition within the area is minimized because few other plants can grow within the clumps. Such clumps may be extremely persistent and capable of regrowth even after complete removal of the foliage by feeding animals or fire, and so the chances of survival of the species are increased.

The vegetative mechanism

Vegetative propagation occurs as a plant reproduces itself from a part of its stem or root. The Canadian pondweed *(Elodea canadensis)*, for example, can reproduce itself from small fragments of its stem. This ability of a cell or group of cells to produce new ones with a function different from that of the parent cells is called cell totipotency.

In higher plants the meristem includes cambium cells. These are small, undifferentiated cells that have not developed a specific function. When properly stimulated they produce the appropriate new part.

Organs of propagation

In higher plants the meristems are also found in special organs, or certain parts of a plant. The organs have two functions—to survive nongrowing periods such as winter or dry seasons, and propagation. Organs of propagation can be underground food or water stores, but are also the aboveground parts of plants.

Many perennial plants grow from a persistent stem base, such as lupines *(Lupinus* spp.). As the plant ages, the stem base increases in diameter, and the older center portions die

Flowering plants: Vegetative propagation

off, leaving the newer outside parts as independent plants.

Some plants develop thick horizontal underground stems called rhizomes, such as irises (*Iris* spp.), Kentucky bluegrass *(Poa pratensis)*, and some ferns, such as bracken (*Pteridium* sp.). Growth is from a terminal bud, but branches also form at intervals, each of which produces a new root. The older parts of the rhizome then die off, leaving the branches independent. Rhizomes produce photosynthetic leaves that grow up through the soil. These aerial shoots are short-lived and die back after one season. The rhizome then overwinters on its food reserves, allowing the plant to survive.

Another mechanism of vegetative reproduction is the production of lateral stems or runners that periodically produce new plants. Strawberries (*Frageria* spp.) reproduce in this way.

The flattened compact base of a stem, which grows underground, is called a corm and is found in plants such as crocuses (*Crocus* spp.). These stems reproduce at their tips on the surface where buds produce new corms directly; but in other plants, such as *Montbretia*, the rhizomes underground may develop corms at their tips.

Tubers are the underground swollen parts of plants that serve as food and water stores. They may become separated by the death of the parent plant and grow into a new plant, the shoots and roots of which may develop tubers and may be tuberous. Root tubers may involve the whole root, as in some orchids, or just the root tip, as in dahlias (*Dahlia* spp.). Roots themselves may generate stems if they become damaged or separated from the stem.

Bulbs, which can be likened to underground buds, frequently produce new bulbs by the growth of an axillary bud within the parent bulb. Bulb scales are fleshy leaves found on bulbs such as onions (*Allium* spp.). These scales store sugar or starch for the plant's future use.

Strawberries (*Fragaria* spp.) propagate by sending out runners from the main stem. These shoots root in the ground and form new plants. The plant has flowers and can reproduce sexually, and they develop into fruit, which may be eaten by animals, though it may be some time before the seeds are excreted and can germinate.

In the potato *(Solanum tuberosum)* tubers form at the end of rhizomes that develop from the base of the stem. Shoots from the buds, or "eyes," of an underground tuber also become tuberous themselves. These tubers are underground swollen stems and can produce new stems, leaves, and roots with new tubers.

The underground stem, or rhizome, of the cowslip (*Primula veris*) acts as a food store throughout winter, but also develops new stems in the spring that bear flowers. After a few years the oldest section of the rhizome dies, and the plants are separated.

The swollen stem base (corm) of the crocus (*Crocus* sp.) is a food source for the plant during winter. It lasts one growing season. A new corm is formed by a thickening of the base of the new stem, just above the old corm.

A tulip (*Tulipa* spp.), *left*, can reproduce by a bulb as well as by sexual reproduction. Bulbs differ from corms in that they are formed from layers of scale leaves that develop underground around the stem. The stem enlarges at its base to form the new bulb and the outermost scale leaves die off.

Monocotyledon

Dicotyledon

Monocotyledonous leaves *(above)* usually sheathe the stem of the plant and have parallel veins but do not have a distinct central vein or midrib. The bird-of-paradise flower *(Strelitzia reginae), right,* which is native to southern Africa, is an exceptional example because its blue-green leaves sometimes have a red midrib.

Monocotyledons

Monocotyledons belong to one of the two groups into which the flowering plants (angiosperms) are divided. They are so called because their seedlings have only one seed leaf, or cotyledon, compared with the two seed leaves of the dicotyledons. Monocotyledons are sometimes considered to be more advanced than dicotyledons. They were originally thought to be derived from dicotyledons. Fossil leaves of both groups, however, have been found in rocks of the same age; the earliest that have been positively identified come from the Lower Cretaceous epoch, making them about 130 million years old. As these plants can be placed in families living today, it is reasonable to assume that they must have evolved at a much earlier time. It is more probable that they both evolved from the same unknown seeded ancestor some time before or during the Jurassic period, from 180 million to 130 million years ago.

Monocotyledons are classified into 9 orders and about 30 families and make up about one-fourth of all flowering plants. They range from small plants, such as snowdrops *(Galanthus nivalis)*, to trees, such as the date palm *(Phoenix dactylifera)*, and include the grasses (Gramineae) and the orchids (Orchidaceae). Monocotyledons grow throughout the world, and many are popular garden plants.

One group, the bromeliads (Bromeliaceae), the group to which the pineapple belongs, mostly exists as epiphytes. They grow upon the branches of other trees—not parasitizing them, merely using them as support. They have aerial roots that hang down and soak up the moisture of the humid forest air.

Grasses are particularly successful. Grasslands will develop wherever there is little scrub or forest cover, provided there is enough moisture.

Stems and roots

The stem structure of monocotyledons is simpler than that of dicotyledons. It is made up of a large number of small parallel, vascular bundles that are closed and are scattered within the stem tissue. Unlike dicotyledons, no true thickening of the stem takes place, although in a few genera dividing cells (cambium) differentiate in the outer part of the stem, forming additional vascular bundles and tissues in which they are embedded.

The core of the trunk of a monocotyledonous "tree" is usually a spongy, fibrous mass of tissue rather than hard wood. The trunks of palms (Palmae) are an exception, their cores being very hard. There are only a few monocotyledon families with species that grow as shrubs or trees. Most belong to the palm family; other woody monocotyledons include

Monocotyledons consist of about 60,000 species of flowering plants. The flower parts are usually arranged in threes and can show a great variety of form from species to species.

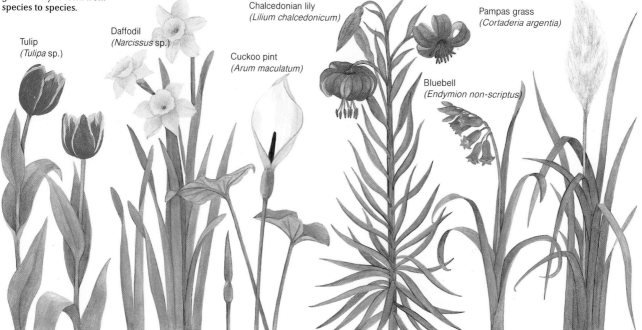

Tulip
(Tulipa sp.)

Daffodil
(Narcissus sp.)

Chalcedonian lily
(Lilium chalcedonicum)

Cuckoo pint
(Arum maculatum)

Pampas grass
(Cortaderia argentia)

Bluebell
(Endymion non-scriptus)

Flowering plants: Monocotyledons

screw-pines (*Pandanus* spp.), dragon trees (*Dracaena* spp.), yuccas (*Yucca* spp.) and some of the aloes (*Aloe* spp.). The banana (*Musa* sp.) is not a true tree because the "stem" is formed out of tightly rolled leaf sheaths and is not woody.

In many monocotyledons, the largest area of growth is underground. The stems grow as rhizomes, corms, bulbs, and tubers, and the aerial projections are only lateral branches or flowering shoots. The roots of monocotyledons differ widely from those of dicotyledons. The primary root, which in a dicotyledon would develop as a taproot, rarely passes the seedling stage before it aborts. In the grasses, the main root disappears soon after germination. The root system of monocotyledons develops from roots growing from the stem. These are simple in structure and do not show any form of secondary thickening. In most species, roots form a fibrous mass, but some large trees, such as several of the palms, have a relatively small root system compared with dicotyledon trees. In some orchids (Orchidaceae), the roots do not branch but remain as simple linear organs. In monocotyledons that have underground stems, such as rhizomes or corms, the roots tend to be annual, dying off at the end of the growing season to be replaced by new ones the following year.

Leaves

The most common type of monocotyledonous leaves are linear or sword-shaped. They have parallel veins—a characteristic of monocotyledons. There are, however, some monocotyledons that have cordate (heart-shaped), ovate, or arrow-shaped leaves, and they generally have a network of veins (reticulate) or ladder-like veined leaves (scalariform). These types of leaves are found mostly in the water plantain family (Alismataceae), yams (Dioscoreaceae), arums (Araceae), and the smilax family (Smilacaceae). The leaves may grow very large, those of the banana (*Musa paradisiaca*) reaching 6 to 10 feet (2 to 3 meters) in length. Generally, the base of the leaf is sheathed, and the leaf stalk (petiole) is frequently absent. But in palms, the petiole is extremely well developed and supports the huge, spreading, hand-shaped (palmate) or feather-shaped (pinnate) leaf.

The leaves of most monocotyledons sheathe the stem and grow alternately on it. The leaf sheath is usually thick and in palms clasps the stem to give the trunk its distinctive ridged appearance. The leaves of aquatic monocotyledons are generally ovate, but the emergent marsh plants may show a variety of leaf shapes. In the arrow-head (*Sagittaria sagittifolia*) the submerged leaves are linear, the floating leaves are ovate, and the aerial leaves are arrow-shaped.

Flowers

In monocotyledon flowers, there are usually five whorls of floral organs: two whorls of perianth segments (sepals and petals), two whorls of stamens, and one whorl of carpels. There are normally three parts to each whorl compared with a varied number in dicotyledons. In many species, especially those of the lily family (Liliaceae) and iris family (Iridaceae),

Date palms (*Phoenix dactylifera*) are monocotyledons that grow in the form of trees. Their stems are like true trunks in that they contain hard woody tissue although the ridges on the stem's surface are not bark but the remnants of dead leaves. They may live for up to 200 years and grow throughout northern Africa and the Middle East.

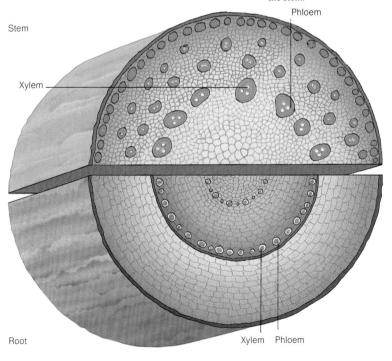

Xylem and phloem vessels in monocotyledons are grouped in closed vascular bundles, arranged centrally in the root but scattered in the stem.

Flowering plants: Monocotyledons

The fly orchid *(Ophrys insectifera)* has flowers that resemble the female tachinid fly. They have "eyes," "feelers," and a blue band of color that looks like the shine on the insect's wings. In addition, the plant is synchronized to bloom at the same time as the male tachinid fly emerges from hibernation. The flowers attract the male fly and are pollinated by it.

the perianth whorls are brightly colored, and flowers in these groups tend to be radially symmetrical (actinomorphic), although there are exceptions. Gladioli (*Gladiolus* spp.) have unequal (zygomorphic) flowers, as do bananas, gingers (Zingiberaceae), cannas (Cannaceae), and orchids.

Many species of monocotyledons show a reduction in flower parts. The flowers of the sweet flag *(Acorus calmus)* have perianth whorls, but these are so reduced as to be inconspicuous; the other flower parts are typically arranged in threes. No perianth whorls are found in the bog arum *(Calla palustris)*, and the flowers are surrounded by a white sheath or spathe (present also in the sweet flag but it is green and does not surround the flower). The number of stamens in the cuckoo pint *(Arum maculatum)* are reduced to three or four, and each ovary contains only one ovule.

There are no perianth whorls, but the flowers are again enclosed in a spathe.

In aquatic plants there is a greater reduction in flower parts. Duckweeds (*Lemna* spp.) have minute flowers that float on the surface of the water. Each flower comprises a spathe surrounding a single stamen. Species of eelgrass (*Zostera* sp.) have further reduced flowers; the male has only one anther and the female one carpel.

Pollination

Monocotyledons are pollinated in a number of ways. Zygomorphic flowers are generally pollinated by animals, such as those of bananas, which rely on bats and birds for the distribution of pollen. The orchids, however, are pollinated by insects, and their flowers show a wide range of adaptations to ensure that the correct insects are attracted to them. In some orchids, such as *Cymbidium*, the flowers twist through 180° as they develop so that they hang upside down when mature, providing a landing place for insects.

Many aquatic monocotyledons are pollinated by water or wind. Canadian pondweed *(Elodea canadensis)* has female flowers that grow at the surface of the water and are pollinated by the submerged male flowers, which break off and float to the surface. In other species of pondweed, submerged flowers are pollinated by water or, if the flowers float on the surface, their pollen is distributed by the wind.

Germination and seeds

The food supply of monocotyledon seeds is generally in the form of endosperm, which is nutritive tissue formed from the embryo sac. The storage material in the seed may be starch or oil. In some families, notably the orchids, the seeds have neither endosperm nor other stored food because they are too small. In this case, the seeds rely on symbiotic fungi, which

Wheat is the staple diet in many countries and is cultivated on a large scale in the United States, Canada, and Europe. The grain is ground into flour, which is used to make bread and pasta.

Flowering plants: Monocotyledons

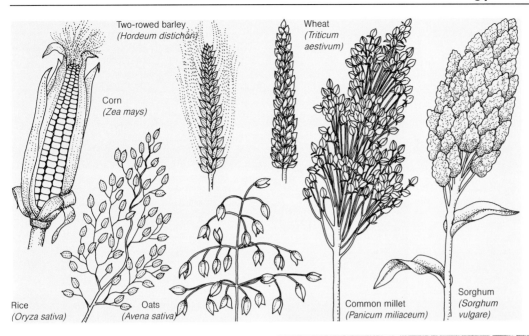

Cereals have provided humans with food for thousands of years. Originally derived from wild grasses, they have been selectively bred to such an extent that some species bear little relation to their ancestors. Grains occupy more than half of the world's harvested land. More than 1.8 billion tons of grain are produced each year. The United States produces about two-fifths of the world's corn, and China more than one-third the rice.

enter their tissues and supply the seedlings with the necessary nutrients during the early stages of their development. During the germination of a monocotyledonous seed, the radicle (embryo root) emerges from the lower part of the seed followed by a bud called the plumule (upper embryo shoot), which is surrounded by the cotyledon. In some species, the cotyledon remains in the seed where the plumule may be surrounded by a special sheath. In grasses, the plumule is also surrounded by a sheath (coleoptile), a similar sheath (coleorhiza) enclosing the radicle. The plumule develops into an aerial shoot, which is usually herbaceous and which may reach a considerable height.

Many different types of fruit are found in monocotyledon families. In the lily family, plants such as tulips (*Tulipa* spp.) bear capsules containing a number of seeds, whereas Solomon's seal (*Polygonatum* sp.) carries its seeds in juicy berries. The sedges (Cyperaceae) produce fruit in the form of nutlets, and the palm tree *Cocos nucifera* bears one-seeded drupes known as coconuts. Flowering rushes (Butomaceae) contain their seeds in a follicle.

Plant products

Man uses monocotyledons for many purposes but by far the most important is their cultivation as a food source. Cereals are grown throughout the world, and their grain provides the staple diet in many countries. Wheat (*Triticum* spp.), rye (*Secale cereale*), barley (*Hordeum* spp.), oats (*Avena sativa*), millet (such as *Panicum* spp., *Pennisetum* spp., *Setaria* sp.), corn (*Zea mays*), rice (*Oryza sativa*), and sorghum (*Sorghum* spp.) are notable cereals. Bamboos are grasses whose woody stems have been used for construction for centuries, as have rattans, the flexible stems of climbing palms. The lily family contains many edible species, including onions (*Allium cepa*), leeks (*A. porrum*), garlic (*A. sativum*), and shallots (*A. ascalonicum*). In parts of Africa and Asia, the root tubers of yams provide an important food source. Some trees of the palm family produce edible fruits, the more familiar of which include the coconut and the date.

Other palms provide valuable fibers. These are extracted from the leaves and are used for brushes and brooms. Another fiber comes from the outer husk of the coconut and is used for matting.

A number of monocotyledon species are cultivated for the fiber that can be extracted from them. Sisal is a strong fiber found in the leaves of some agaves (*Agave* spp.) and when twisted can be made into rope. Other notable fibers obtained from monocotyledons include New Zealand flax (*Phormium tenax*), bowstring hemp (*Sansevieria* spp.), and yuccas.

Many monocotyledons are grown as garden plants, particularly members of the Amaryllidaceae family, including snowdrops and daffodils (*Narcissus* spp.). Irises, freesias, gladioli, and crocuses all belong to the iris family and are popular garden plants.

Bamboos are classified as grasses, but unlike most grasses, some species may reach up to 120 feet (37 meters) in height. They thrive in subtropical to mild temperate climates and are found mainly in southeastern Asia. Young shoots of some species can be eaten, and the hollow stems of mature plants are used for light construction.

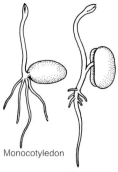

Dicotyledons are characterized by the presence of two seed leaves (cotyledons) in the embryo contained within the seed. In contrast, most monocotyledons have only one seed leaf.

Dicotyledons

The angiosperms are a large group of plants that, unlike gymnosperms, bear flowers and have seeds that are completely enclosed. In many modern systems of plant classification, the angiosperms (subphylum Magnoliophytina) are divided again into two classes: the dicotyledons (Magnoliopsida) and the monocotyledons (Liliopsida). (This classification, however, may vary. *See The World Book Encyclopedia* article FLOWER [How flowers are named and classified].) The major basis of these groups, and the one that gives them their names, is their number of seed leaves: dicotyledons have two seed leaves and monocotyledons have one.

There are more than 190,000 species of dicotyledons and all of them, from garden flowers, such as hollyhocks (*Althaea* spp.), to complex trees, such as oaks (*Quercus* spp.), have within their seeds two seed leaves or cotyledons. These may remain underground during the germination process or appear above ground. When visible, the cotyledons have a simple, usually rounded shape.

Leaves, shoots, and roots

Apart from two seed leaves, dicotyledons have other distinctive anatomical features. Their true leaves (those other than the seed leaves) have veins, usually arranged in a netlike pattern with a distinctive central vein, or midrib, and the base of the leaf usually tapers to a point.

The stems of dicotyledons also have characteristics that typify the group: the water-conducting cells (xylem) and those that transport dissolved food substances (phloem) are grouped in open vascular bundles arranged around the perimeter of the stem. Between the xylem and phloem is a growing layer (cambium), which produces new xylem and phloem cells in woody plants (shrubs and trees).

Another distinctive feature of most dicotyledon stems is that, as their complement of xylem and phloem cells increases through the activity of the cambium, the stem becomes toughened or thickened. If this process—known as secondary thickening—continues, the stem becomes woody.

Like their stems, the roots of most dicotyledons also contain xylem, phloem, and a cambium capable of creating secondary thickening. Whereas the water-conducting xylem tissues of monocotyledons are commonly arranged in a ring, those of a dicotyledon root take on a characteristic cross or star shape.

Flowers

Monocotyledon flowers, such as tulips (*Tulipa* spp.) and lilies (*Liliaceae*), have their parts arranged in threes, whereas those of the dicotyledons have a different, more varied numerical formula of construction. Most dicotyledons have flowers with their parts—including petals, sepals, the pollen-producing stamens and the pollen-receiving stigmas, and the parts of the ovary in which embryo seeds are contained—grouped in fives or fours. In the dicotyledon flowers, which are thought to have evolved first and which are therefore the most "primitive," the parts are numerous and are arranged in whorls, not joined together. The composites—that is, daisies and their relations—have the most specialized dicotyledon flowers, made up of hundreds of tiny florets.

Dicotyledon flowers are structurally quite varied, the differences in petal structure being one of the most significant features. Plants whose flowers have free (unjoined) petals, such as roses (Rosaceae) and buttercups (*Ranunculus* spp.), and those with no petals, such as those of many trees, including the birch (*Betula* sp.), are placed together in one group, the

Magnolias (*Magnolia* spp., *right*) are considered to be among the most "primitive" (least highly evolved) genera of dicotyledons. The parts of the flower are numerous, unjoined, and arranged in whorls. The wallflower (*Cheiranthus* sp., *far right*) is intermediate in evolutionary development, having flowers with four unjoined petals, four sepals, six stamens, and one stigma.

Flowering plants: Dicotyledons

Archichlamydeae. Those with petals completely or partly fused, such as gentians (*Gentiana* spp.), bluebells (*Campanula* spp.), and heathers (*Erica* spp.), belong to a second group called the Metachlamydeae.

Distribution and diversity

Dicotyledons have evolved to occupy all kinds of habitats, from tropical forests to high mountains, from deserts to fresh or salt water. Only being blanketed with a permanent covering of snow and ice acts as a total deterrent to species of dicotyledons.

In their life styles, dicotyledons show an equal diversity. At the extremes, for example, they may be "insect-eaters," such as the sundew (*Drosera* sp.), or pale parasites, for example, the dodder (*Cuscuta* sp.), which absorbs its nutrients from the stems of other plants.

Dicotyledons may be perennials, biennials, or annuals. In some harsh environments, annuals grow, bloom, and die in the space of a few weeks. The perennial species may be herbaceous—dying down to ground level each year—or woody, with a permanent growth structure above the soil level.

The dicotyledons include many of the most prolific wood- and food-producing plants. All the world's hardwoods, from oak to teak, come from dicotyledons. Of the many plant families that provide food for human beings and their domestic animals, the following dicotyledons are among the most significant: the Cruciferae, which includes all of the cabbage family; the Solanaceae, to which potatoes and tomatoes belong; the Rosaceae, from which come apples, pears, and other soft fruits; the Compositae, whose members include the chicories and lettuces; the Cucurbitaceae, the family of melons, marrows, and cucumbers; and the Leguminosae, from which come the beans and peas.

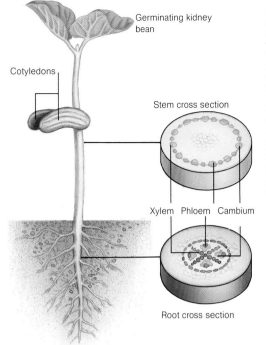

The two cotyledons (seed leaves) of a seedling kidney bean appear above the ground. A cross section of the bean's stem reveals, as in most dicotyledons, vascular bundles (groups of xylem and phloem cells) arranged in a ring around the perimeter of the stem. These cells are separated by a growing layer called the cambium. The root of a typical dicotyledon has its xylem and phloem cells grouped at the center—the xylem arranged in a cross or star shape, with the phloem between it.

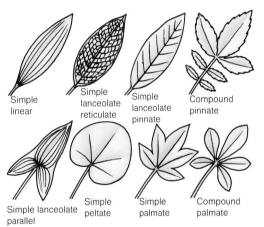

Leaves of dicotyledons vary widely in shape, size, and complexity. Undivided (simple) leaves include the linear campion, the peltate nasturtium, and the palmate maple leaf. Typical compound leaves (made up of several leaflets) include the pinnate vetch leaf and the palmate horse chestnut leaf.

The inflorescence of the daisy (*Bellis* sp.) consists of many florets. In such a highly evolved flower, only the central disk florets are capable of forming seeds. The purpose of the outer, petallike florets is to attract pollinating insects and birds.

Herbaceous plants

The term herbaceous, in its botanical sense, is applied to plants that usually have no woody parts. In general, the leaves and stems of these plants die back at the end of the growing season. Perennial herbaceous plants overwinter by means of underground storage organs; annual herbaceous plants usually survive winter by means of seeds.

Herbaceous plants do, however, show extremely varied forms. One of the smallest is the aquatic species of duckweed *Wolffia arrhiza,* which is a minute green plant that floats on the surface of fresh water. It has no true roots or stem and is about 0.5 to 1.0 millimeter across. Conversely, herbaceous plants can grow to a large size; the taro *(Colocasia esculenta)* produces leaves 3 feet (91 centimeters) or more in length. Other herbaceous forms include prostrate creepers, such as the creeping jenny *(Lysimachia nummularia),* and matlike plants, such as the pearlwort *(Sagina* sp.).

Perennials

There are three basic types of herbaceous plant, categorized according to their life cycles: perennials, biennials, and annuals. Herbaceous perennials live for a number of years, surviving from one growing season to another. In temperate climates, perennials must survive a winter between two summers, and in tropical areas they may have to live through a dry season between two wet seasons. This survival is generally achieved by means of a surface-dwelling or underground organ (which is also by a means of vegetative reproduction). These organs lie dormant during the cold season and produce new aerial parts each year, which die back at the end of the growing season.

The life span of perennials varies from species to species. Some plants are generally short-lived and others, such as the peonies *(Paeonia* spp.), may persist for many years. Many familiar garden flowers are herbaceous perennials and include such plants as Michaelmas daisies *(Aster* spp.), pampas grass *(Cortaderia selloana),* dahlias *(Dahlia* spp.), chrysanthemums *(Chrysanthemum* spp.), and numerous plants that have bulbs, corms, and tubers.

Many terrestrial and epiphytic tropical plants—the bromeliads (Bromeliaceae), for example—are also herbaceous perennials although their aerial parts persist from one growing season to another.

Biennials

As their name suggests, biennials live for two years. In the first year, the seedling grows into a rosette of leaves attached to an underground storage organ. This storage organ becomes swollen with food at the end of the growing season and, in the following year, produces a flowering stem with a few sparse leaves. After

Herbaceous borders are a common feature in cultivated gardens. All garden flowers originated in the wild, but many have been selectively bred to form hybrids. The longer-lasting perennials are generally planted at the back of a herbaceous border where they are allowed to grow tall. Biennials and annuals are usually smaller plants and are planted at the front of the border where they can be easily replaced.

A weed is any plant that grows where it is not wanted. Dandelions are successful weeds because they are fast growers and can survive even when most of their leaves have been removed.

Flowering plants: Herbaceous plants

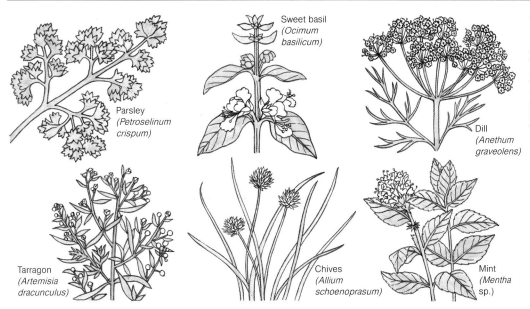

Culinary herbs, as their name suggests, are mostly herbaceous perennials and annuals used in cooking throughout the world. Sweet basil, which is native to India and Iran, is often used in tomato-based recipes, whereas dill is a European herb used in sauces and salads.

Parsley (*Petroselinum crispum*)
Sweet basil (*Ocimum basilicum*)
Dill (*Anethum graveolens*)
Tarragon (*Artemisia dracunculus*)
Chives (*Allium schoenoprasum*)
Mint (*Mentha* sp.)

the seeds have been dispersed, the whole plant—including the underground storage organ—dies away.

Occasionally an axillary bud at the base of the stem remains; this will develop as another basal rosette for the following year, with the result that seeds are produced every year rather than every two years. The common foxglove *(Digitalis purpurea)* has axillary buds that persist regularly, as does the garden hollyhock *(Althaea rosea)*.

Many crop species are biennials and include carrots *(Daucus carota)*, parsnips *(Pastinaca sativa)*, and some species of cabbage.

Annuals

Each year, a new crop of annual plants grows from the previous season's seeds. The germinating seedlings mature, flower, scatter their seeds, and die within one growing season. Some of the seeds may survive in a dormant state for several years, until conditions are right for germination. Many culinary herbs are herbaceous annuals and include basil *(Ocimum basilicum)*, caraway *(Carum carvi)*, coriander *(Coriandrum sativum)*, and dill *(Anethum graveolens)*. Annuals grown as crops include peas *(Pisum sativum)*, lentils *(Lens culinaris)*, chick peas *(Cicer arietinum)*, and all the grain crops.

Similar in character to the above annuals are shorter-lived annuals. These plants have an even shorter life cycle. They manage to compress the period from germination through seed dispersal to death of the parent plant into a few weeks. If conditions are favorable, several life cycles may take place in the space of one growing season. This group of plants includes many weeds and desert plants. Groundsel *(Senecio vulgaris)* and, in Europe, shepherd's purse *(Capsella bursapastoris)* are typical ephemerals and, as a result, are successful weeds.

Stem structure

All flowering plants, both monocotyledons and dicotyledons, can be separated into two groups on the basis of their stem support structure. Plants such as trees and shrubs have persistent woody aerial parts that survive for a number of years. Nonwoody or herbaceous plants do not have persistent aerial parts although the stems of some species, such as goldenrod (*Solidago* sp.) and sunflower (*Helianthus* sp.), may show some secondary thickening.

The stems of small and short-lived herbaceous plants depend mostly for their support on living tissues with thickened walls (collenchyma) found on the outer areas of the stem and on the veins of the leaf. All herbaceous plants, however, have some strengthening tissues associated with the vascular bundles and, in most species, there is some production of secondary tissue within the bundles. Other herbaceous plants, such as the garden hollyhock, have dead, thick-walled tissue (sclerenchyma), which often contains lignin and helps to support the stem. The distribution of sclerenchyma is often related to the areas of the

Passion flowers (*Passiflora* spp.) are herbaceous vines that climb with the help of tendrils. Many species are cultivated for their brightly colored flowers and their edible fruits. This plant grows wild in the Bolivian Andes at an altitude of about 6,000 feet (1,800 meters).

Flowering plants: Herbaceous plants

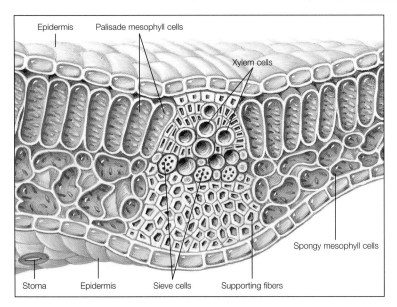

A transverse section of a leaf shows a central vein (vascular bundle) with supporting fiber cells with thick walls below and bundle sheath cells at the sides and top. Leaves are kept rigid by their veins and the palisade and spongy mesophyll cells that are full of water (turgid).

plant that are subjected to mechanical stress.

Roots and storage parts

Herbaceous plants have many different types of root system and underground storage parts. Fibrous root systems are common in herbaceous perennials and grasses, whereas taproots are found in plants such as carrot and dandelion (*Taraxacum* spp.).

In some parasitic species, the seed germinates and the root grows underground without producing an aerial shoot, except at intervals when the plant is ready to reproduce. The curious-looking bird's nest orchid *(Neottia nidus-avis)* and Indian pipe orchid *(Monotropa uniflora)* produce an aerial stem only when the plant is flowering, with simple yellow or pink leaves and flowers that hang from the end of a branch. Other species of orchid grow in a similar way, such as the coral root (*Corallorrhiza* sp.) and ghost orchids (*Epipogium* spp.).

Certain species have no roots and produce only shoots. In the case of such species, the plant's stem takes over the water- and mineral-absorbing function of the roots. Typical of these species are the aquatic and terrestrial bladderworts (*Utricularia* spp.), which have modified leaves—known as rhizophylls—that behave like roots.

Rhizomes make up the permanent axis of many herbaceous perennials and are swollen stems (usually partially or totally underground). Whereas aerial shoots may die down at the end of each growing season, rhizomes survive to produce new shoots the following year. Similarly, bulbs are the swollen extensions of perennial stems and are formed from a mass of modified leaf bases. Some perennials rely on underground storage parts called corms, which are also swollen stem bases.

Types of leaves and tendrils

Herbaceous leaves show a great variety of shapes. The most common leaf form is sword-shaped or linear, as produced by the irises (*Iris* spp.), grasses (Graminaceae), and other monocotyledons. Leaf edges can be smooth or have simple serrations or they may have deep regular or irregular incisions—for example, those of the dandelion. A further development of this form is the pinnate leaf, which is a characteristic of poppies (*Papaver* spp.) and many umbellifers (Umbelliferae).

Other sorts of simple leaf form include palmate (as in geraniums, *Geranium* spp.), spear-shaped (as in some arums, *Arum* spp.), heart-shaped (as in the lesser celandine, *Ranunculus ficaria*), and circular (as in penny-wort, *Hydrocotyle* spp.). Leaves like those of the lupine (*Lupinus* spp.) are said to be digitate, and those of hellebores (*Helleborus* spp.) are pedate.

Carnivorous herbaceous plants have a variety of unusual leaf forms that are usually adaptations to their capturing insects. Pitcher plants

The "root" vegetables that we eat are not always true roots in the botanical sense. Potatoes are the swollen extensions of underground stems (tubers) of herbaceous plants, whereas onions are modified storage leaves (bulbs) of herbaceous monocotyledons. Carrots and radishes, however, are the swollen roots of herbaceous plants.

(*Sarracenia purpurea*), as their name suggests, have funnel-like leaves that trap insects that are subsequently digested by enzymes. The leaves of sundews (*Drosera* spp.) are covered with sticky glandular hairs that trap insects. The Venus's-flytrap (*Dionaea muscipula*), another carnivorous plant, has a spring mechanism that allows the leaves to snap shut when the trigger hairs on their surface are touched, thus trapping any insect that may be on the leaves.

Some herbaceous plants have tendrils (modified leaves) for climbing—for example, the legumes (Papilionaceae). In some climbing herbaceous plants, tendrils are entirely separate parts of the plant. The cucumber family (Cucurbitaceae) has tendrils that grow from the leaf base, and in white bryony (*Bryonia dioica*), half of each tendril twists in one direction and half twists in the other direction so as to pull the plant closer to its support. In certain species, it is the plant's branches that are the means of support; in others, it is the stem itself that twines, such as bindweed (*Convolvulus* sp.).

Some species are scramblers, supporting themselves on other, taller vegetation by means of hooks and stiff hairs—for example, bedstraw (*Galium aparine*). Species such as dodder (*Cuscuta* sp.) are parasitic climbers that obtain their nutrients from the host's stem. Others are parasitic on the roots of plants, such as the broomrape (*Lathraea clandestina*). These plants may have no aerial shoots, and their flowers are produced at the soil surface.

Growth rate

Herbaceous plants can produce a phenomenal amount of growth each year. Plants such as rhubarb (*Rheum rhaponticum*) and butterbur (*Petasites hybridus*) produce leaves that often measure 3 feet (91 centimeters) or more in diameter, and some temperate species, such as gunnera (*Gunnera manicata*), can grow leaves that are twice this size.

Perennial climbers such as bindweed (*Convolvulus* sp.) produce many feet of growth each year and are unpopular with gardeners, as is morning glory (*Ipomoea* sp.). Many biennials and perennials produce stems up to 6 feet (1.8 meters) tall with a similar-sized spread of side-shoots—for example, the teasel and cartwheel flower (*Heracleum mantegazzianum*). These species have perennating organs that allow them to start growing early in spring. Himalayan balsam (*Impatiens glandulifera*), which is an annual plant, produces as much growth starting from seed each year as some perennial species. Its stem may measure up to 2 inches (5 centimeters) at the base and is produced in about 5 months.

Cartwheel flower (*Heracleum mantegazzianum*) is a fast-growing perennial that can reach up to 13 feet (4 meters) in height in one growing season. It is native to the Caucasus but is grown elsewhere as an ornamental plant.

Annuals are plants that may condense their growing season into a few weeks. In Central Australia, where rain is scarce, plants take advantage of a summer shower to germinate, flower, and scatter their seeds.

Shrubs

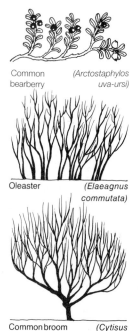

Shrubs grow in many shapes and sizes, ranging from climbers to small trees. The bearberry, oleaster, and broom *(shown in silhouette, above)* illustrate three common ways in which shrubs grow: they creep along the ground, branch from just beneath the surface of the ground, or branch from a central "trunk," like trees.

Common bearberry *(Arctostaphylos uva-ursi)*
Oleaster *(Elaeagnus commutata)*
Common broom *(Cytisus scoparius)*

Shrubs grow naturally in many of the world's floras, from savanna to shrub, from the tundra to the tropical rain forest. Cultivated shrubs are common in gardens throughout the world.

There is no clear line of distinction between shrubs and trees or between semi-shrubs and subshrubs, but certain generalizations can be made. Plants that grow in the form of shrubs are perennials with woody stems but, unlike trees, have little or no trunk. They may grow as bushes or, in some cases, as climbers. Shrubs branch from near or just below the ground and reach a maximum height at maturity of between 1.5 and 16 feet (0.5 and 5 meters). In addition, they differ from trees in the way in which they allow their lateral buds to grow. A tree prevents these buds from growing except at special branch points, whereas a shrub lets as many grow as is practical. Some shrubs are evergreen and some are deciduous, but almost all are angiosperms—that is, most reproduce sexually by means of flowers.

Types of shrubs

Different groups of shrubs have their own particular attributes. Some shrubs thrive in hot wet environments such as the tropical forests. Most of these are evergreen, for example, members of the genus *Philodendron,* which have big, broad leaves—a characteristic that makes them popular house plants.

The tropics and subtropics also support plants that are able to survive wide variations in rainfall. They are deciduous shrubs, which shed their leaves during a cool or dry season. When the new buds grow, at the arrival of warmth and rain, they are initially protected by scales, sometimes leaflike in form. But tropophytic shrubs are not confined only to the warmer parts of the world. They are also the dominant deciduous shrubs of temperate regions and extend as far as the Arctic.

In the warmer parts of the world's temperate regions, conditions favor shrubs with large evergreen leaves, which have a thicker epidermis than some other types of shrubs. The Mediterranean sweet bay *(Laurus nobilis)* is typical, as are the camellias from India, China, and Japan.

Many dwarf species, such as the heathers *(Erica* spp.), are regarded as shrubs because of their woody, branching stems. They inhabit heath and moorland areas and grow to less than 1.5 feet (0.5 meter) tall, extending horizontally because high winds restrict vertical growth. Many leaves grow within the bush where they are protected from grazing animals.

Some groups of shrubs are categorized according to the particular adaptation that allows them to withstand extreme weather conditions, such as heat, drought, and gales. Examples include many of the acacias *(Acacia* spp.), which have small, rigid leaves. Because the leaves are small, water loss is minimized during drought, and their rigidity supports the leaf structure in dry, windy conditions.

The heathers are xerophytic; they conserve water by means of leaves that are rolled up at the edges, thereby sealing off the pores (stomata) through which water loss takes place during transpiration. Thorn shrubs—again including many of the acacias—have spiky leaves, which are reduced to needlelike structures. This adaptation helps to conserve water by presenting a smaller surface area from which evaporation can occur. Switch shrubs, such as the brooms *(Cytisus* spp.), prevent excess transpiration by having leaves of many minute, overlapping scales or in some cases by having no leaves at all. Tola shrubs are typified by their needle- or scalelike resinous leaves.

Two other groups of shrubs are also drought-resistant but work on a different principle: they store water as well as prevent its loss. Succulent shrubs of the salt marshes and salt deserts—halophytes—have fleshy water-retentive leaves borne on their woody stems. Other shrubs, such as some members of the genus *Eucalyptus,* store water in swollen, underground stems.

Gorse *(Ulex europaeus),* below, is a spiny evergreen shrub growing up to 4 feet (1.2 meters) in height. Its seeds are contained in hairy pods, which explode loudly when ripe.

Flowering plants: Shrubs

Colonization

In the succession of plant species that occurs when land is colonized by plants, shrubs tend to follow annuals and herbaceous perennials, such as the grasses, but precede trees. When tree growth does occur, it is often at the expense of shrubs because the trees tend to deprive the shrubs of light and moisture, which they need for survival. In many forest habitats, however, both deciduous and coniferous shrubs benefit from gaps caused by fire or falling trees or from human interference. They colonize quickly, producing a large quantity of seed shortly after they start to grow. The cropping of forests, for example, leads to the establishment of shrubs, which dominate the habitat until succeeded by trees that have grown naturally from seeds or that are planted in reforestation programs.

Only in a few areas are shrubs the dominant form of plant life. Most notable of these are the shrublands, which are of two kinds—the chaparral and the scrublands found in continental interiors. Chaparral occurs in coastal regions of California. Similar shrublands occur in other Mediterranean climates, for example, France, the southwestern Cape of Africa, and southern and southwestern Australia. Thorny and succulent shrubs are typical of the scrublands of continental interiors. These areas, which occur, for example, south of the Sahara Desert and in the interior of Australia, are usually bounded by desert.

Yucca plants *(above)* are a genus of the agave family and live in arid areas of the Southwestern United States and in Central and South America. Most yuccas rely on the "yucca" moth for pollination, which is attracted by the smell of the cream-colored flowers.

When land is stripped by fire or human activity, such plants as annuals and herbaceous perennials are quick to colonize it (A). Shrubs eventually establish themselves (B), but when they are mature, shade the herbaceous plants from light. Similarly, after many years, trees come to dominate the land (C) and cut out the light that reaches shrubs and plants on the forest floor.

Climbers

There are many species of flowering plants that have developed the ability to climb up other plants or structures. Their stems are not strong enough to be self-supporting. Climbers redirect some of their energy into producing structures such as tendrils, which assist them in climbing up their vital supports.

Climbers come from a great range of plant families, from the cucumbers (Cucurbitaceae) and grapevines (Vitaceae) through representatives of tropical families such as the bignonia (Bignoniaceae) and passionflowers (Passifloraceae). Nearly all climbers are dicotyledons, with the exception of the rattans, which are monocotyledons of the palm family (Palmae).

In many habitats, and particularly in dense jungles, climbing is the method these plants use to reach the sunlight that they need for photosynthesis. The energy saved by not developing large stems can be used to grow faster, farther, and higher than neighboring plants, which allows them to stay in the light and gain even more energy. Their flowers are then more likely to be produced in an environment that is conducive to fertilization. Many climbers, however, are also able to reproduce vegetatively by means of organs such as underground stems.

The woody vine, wisteria *(Wisteria sinensis)*, is a popular garden plant for climbing up walls, which it does by twining its stems around a support.

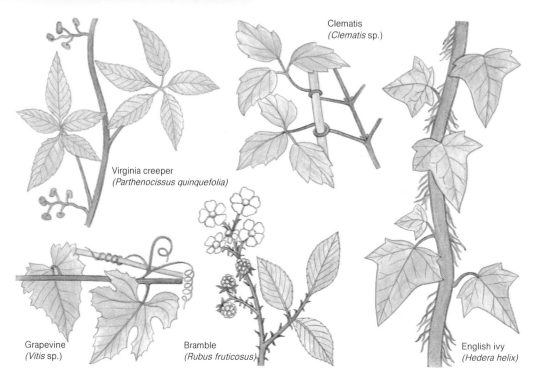

Plants may climb using suckers, as does Virginia creeper, or modified branches that have become tendrils, as on the grapevine. Some plants, for example clematis, wind their whole stem around a support. Adventitious roots on the cablelike stem of ivy attach the plant to a surface. Trailing climbers, such as brambles, hook onto supports with their downward-pointing thorns.

Virginia creeper *(Parthenocissus quinquefolia)*

Clematis *(Clematis sp.)*

Grapevine *(Vitis sp.)*

Bramble *(Rubus fruticosus)*

English ivy *(Hedera helix)*

Means and methods of climbing

Climbing is accomplished by a variety of methods, and some plants possess special organs for this purpose. The simplest way in which a plant climbs is by twining itself around a support that can be either natural—for example, another plant—or artificial. Most twining climbers are left-handers—that is, the direction in which they twine when viewed from above is counterclockwise. A few—notably the hop (*Humulus lupulus*) and the honeysuckle (*Lonicera periclymenum*)—are right-handers.

Shoots, leaves, and even roots of flowering plants may be modified to assist the climbing process. Shoots may bear sharp down-curved spines, often reinforced with woody tissues, such as those of roses (Rosaceae), which hook onto the surrounding vegetation. Special shoot outgrowths may be equipped with suckers like those of the Virginia creeper (*Parthenocissus quinquefolia*). Tendrils, which are generally formed as modified leaves, may be a simple spiral or may be armed with hooks. In some species, the tendril bears a leaflike structure, such as that of a vetch (*Vicia* sp.), which may have a simple tendril in place of the uppermost leaflet. The rapidly climbing ivys and their relations produce special roots from cablelike stems, which anchor the plants so that they can grow up the surface of their support. These roots are borne in groups or rows but only on the shaded side of the stem. If the support is a living substrate, such as a tree, the roots may penetrate it and take up nutrients. This process is not likely to kill the support, but its eventual death may be hastened as a result.

How climbers work

The tissues of flowering climbers are often modified in their arrangement to give maximum strength and flexibility. For this purpose, a large number of sclerenchyma fibers may be contained in the stem. Many climbers, particularly the woody jungle lianas, have stems in which rays of phloem-containing pith run through the woody xylem. Liana stems do not cling to their supports, however, but attach themselves more than 150 feet (46 meters) above the ground, with the "trunk" left to hang free.

The xylem vessels contained in the woody tissues of the tallest climbers are very large—in the rattans, for example, they may be 10 to 20 feet (3 to 6 meters) long. These and similar vessels conduct water up the climber at a spectacular rate, sometimes as fast as 6 feet (1.8 meters) per minute. If a section is cut out of a tropical climber and then held upside down, the water in it pours out.

Climbing movements

Apart from their physical structure, climbing plants are largely reliant on innate physiological mechanisms. These mechanisms depend on the plant's ability to sense light, gravity, touch, and temperature, and to respond accordingly by a movement, which generally involves growth. Like those of most plants, the

The wild hop (*Humulus* sp.) climbs by twining its stem around anything it can. It is a right-handed climber, climbing in a clockwise direction, unlike most other twining climbers.

The tropical climber balsam pear (*Momordica charantia*) uses spiraling tendrils to attach itself to other plants. This annual plant is a member of the cucumber family, Cucurbitaceae, and grows in the Old World tropics. Its orange fruit splits to reveal an edible pulp and to disperse the seeds.

shoots of climbers respond positively to light (phototropism) by growing toward it and negatively to gravity by growing away from it. The roots behave in an exactly opposite way. There are exceptions, however, such as the wild grape (*Vitis* sp.), the tendrils of which are negatively phototropic. They end in disks which attach themselves to their support.

As a shoot grows it can be seen to move about in circular arcs. This natural tendency seems to arise from an inner control on the part of the plant—the stimulus is received at sites in the epidermal cells and is rapidly conducted across the whole tendril. Shoots are also affected by sensitivity to touch. Once a support has been touched by the encircling shoot, a growth spurt is triggered on the surface of the shoot directly opposite the point of contact. This surface grows longer and more rapidly than the side of contact, causing the shoot to bend inwards toward the contact and grow around the support. The growth is so rapid that it can be seen over a period of a few hours. Once begun, the shoot continues to grow in the same direction.

Trees

Most tree species are found among the gymnosperms and angiosperms, with more species occurring in the latter group. Most are dicotyledons, but a few are monocotyledons, such as palms (Palmae), Joshua trees *(Yucca brevifolia)*, which are found in desert areas, and dracaenas (*Dracaena* spp.).

Botanically, trees are plants that usually have a single woody stem (the trunk, or bole) and a crown with woody branches. They may be only a few feet high, as in the undergrowth of tropical rain forests, or very tall—the Australian mountain ash *(Eucalyptus regnans)* and the California redwoods, for example, may attain a height of more than 300 feet (91 meters).

Only gymnosperms and dicotyledons have proper wood. This substance is a permanent, secondary tissue comprised of cells with lignified walls. Plant cell walls are made mostly of cellulose arranged in microscopic fibrils. In woody plants the cellulose is impregnated with lignin—a material that imparts compressive strength to the cell wall.

Tree growth and form

The apical meristems of a tree—the growing regions at the tips of the trunk, twigs, and roots—are the sources of increase in height of the tree. Auxins (growth hormones) are supplied in greater quantities to the trunk meristems than to the lateral ones, which are often suppressed.

The secondary or lateral meristem, in the shape of a thin ring between the wood and the bark, is the vascular cambium. This meristem is the means by which lateral growth, or thickening, takes place in the tree because it forms new wood cells inward, and bark cells on its outer side. Monocotyledon trees do not have true wood, but numerous vascular bundles set in a matrix of parenchyma, which may become lignified. They either have no secondary thickening (such as palms) or thicken by the formation of extra vascular bundles as do dracaenas, for example.

In some tree species the trunk divides at a low level into two or more trunks. In most trees, a single trunk forms the dominant axis from which lateral branches grow to form a crown, or head, of branches. Those species in which the trunk continues to the tip of the tree as a single axis, with the lateral branches being smaller in size and of secondary importance, have a pyramidal, or excurrent shape, and are described as monopodial (single axis). Trees in which the trunk does not grow to the crown top but is replaced by variously massive limbs, are termed decurrent, or sympodial (many axes), such as oaks (*Quercus* spp.) and basswood (*Tilia* spp.). When young, all trees are monopodial. Few dicotyledons retain this form at maturity, although many gymnosperms do. In monopodial trees where the leaves occur only at the top of the trunk, as in palms, the trees may be described as columnar.

The shape of a tree varies partly according to its species and partly in response to its environment. In a specimen tree—a tree grown in the open in isolation—the branches remain, atypically, from low down on the trunk and may spread out, almost sweeping the ground. When trees grow close together, as in a forest, the lower branches receive little light and soon die. The branches forming the crown at the top of the tree are the only ones that survive and the tree grows having a branch-free trunk.

The influence of the environment on the final shape of a tree is well seen in exposed

Trees vary greatly in shape according to their species.

The growth rings on the trunk of wych elm *(Ulmus glabra)* display the different annual rates of growth. The pale sapwood contains xylem, which transports water, and inner phloem (bark), which conducts food. The inner dark section is the heartwood. The small rings on the edge of the stem belong to a lateral branch.

windy areas, such as mountaintops or near the sea. A tree that manages to grow there develops a distinctive swept-back look with the branches trailing away from the direction of the prevailing wind. As soils in these places tend to be thin or nutrient poor, the tree never grows to such a height as the same species would in a sheltered area with an abundance of rich deep soil.

These shapes can be developed artificially. The art of bonsai, perfected in Japan, involves the development of a shallow root system by growing in a small pot and regular root pruning. This has the same effect as when a tree grows in shallow soils. The branches are also pruned and trained with wires to grow in the required direction, simulating the effect of the wind. By manipulating the tree's environment in this way, the bonsai master can produce miniature versions of any tree form found in the wild.

Annual rings and wood rays

A cross section of many tree trunks reveals a number of close concentric circles that appear to be split into wedges by raylike lines. These circles are annual growth rings and are apparent in trees that grow in temperate climates only. The rings are formed because the cambium in the trunk produces large conducting cells in the spring and early summer and smaller conducting cells in late summer. The highly porous spring wood conducts water rapidly to the new growing shoots, and the less porous summer wood acts as a strengthening tissue. Because growth ceases in the fall in temperate regions there is a distinct boundary between the ring of one year's growth and the next; but this is not so in trees that grow in equatorial climates, where there are no distinct seasons. Some tropical areas have wet and dry seasons which may form distinctive rings in the wood, but they are not always reliable annual indicators.

Not only does the historical analysis of annual rings (dendrochronology) signify the age of a tree with considerable accuracy, it also indicates good and bad growth seasons, which can give a measure of the rainfall at a certain time in the past. Radioactivity levels, the oxygen content of the air, and mineral nutrient levels in the soil, among other things, can be discovered by chemical analysis of a single tree ring.

The rays that divide annual rings into wedges radiate outward from the center of the trunk. They are known as wood rays and are made up of parenchyma cells. As the tree grows in width, the cambium produces new conducting tissues and packing cells of parenchyma. The ray parenchyma cells transport food, waste chemicals, and other materials toward the center of the trunk. The stored food (often starch) can be mobilized in spring when the buds open and is passed into the vessels to be carried to the areas of growth. In spring the sap consists of a solution of up to 8 per cent sucrose. In the sugar maple (Acer saccharum), for example, this "sugar run" is exploited each year when people tap the vessels for the sugar-rich sap.

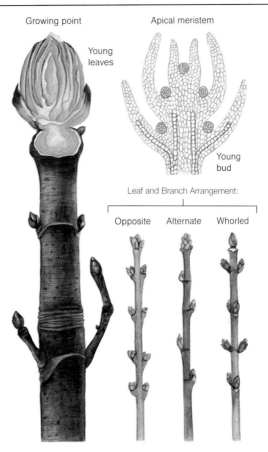

The growing point, or apical meristem, on a stem consists of meristem tissue in which new cells are formed by rapid cell division. Buds form in the inner junction of each leaf with the stem and grow on stems in three ways—opposite each other, alternately opposite, or whorled, spiraling up the stem.

Sapwood and heartwood

A transverse section of dicotyledonous wood reveals two types—a light, outer layer of sapwood and an inner, darker layer (the heartwood). The sapwood has living parenchyma

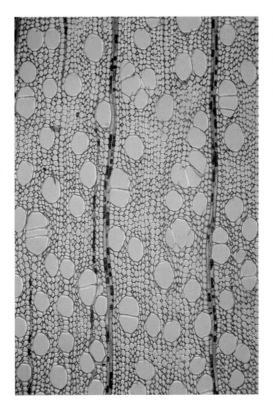

Maple wood is a diffuse-porous hardwood. The term applies to the vessels in the sapwood, which are more or less the same size and evenly distributed throughout the spring and summer wood.

Flowering plants: Trees

A longitudinal section through the sapwood of a beech tree displays the wood fibers and vessels, running vertically, and the rays, running horizontally.

cells and water-conducting xylem cells, tracheids, and vessels, as well as fibers. Vessels occur in angiosperm wood and not in that of gymnosperms. In the latter, the tracheids and fibers are usually not fully differentiated but exist as fiber-tracheids.

The different cells of the xylem and their arrangement give the wood its characteristics—the vessels and fibers are often present in different proportions in spring and summer wood. In some trees—oak, for example—the broad vessels are confined to the spring wood whereas in the summer wood there are numerous fibers and fiber-tracheids. This type of wood is known as ring-porous wood. Diffuse-porous wood, as found in maples (*Acer* spp.) and basswood, has much smaller vessels, which are more evenly distributed throughout the annual rings. A few genera of angiosperms are exceptional in that, like gymnosperms, they have no vessels. This feature is believed to be a primitive one and is found in *Drimys* and *Trochodendron* of tropical forests.

In contrast to the sapwood, the cells of the heartwood are mostly nonconducting and are used for storage except at the boundary with the sapwood where the parenchyma cells store water. The parenchyma cells of the newly formed sapwood usually live for several years. As sapwood ages to form heartwood, its cell walls undergo a chemical change and become darker and denser. Heartwood also con-

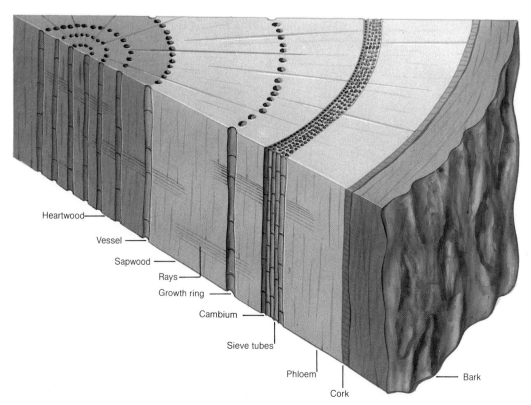

A dicotyledonous tree trunk consists of rings of several compositions. The innermost dead section—the heartwood—is darker and denser than any other part of the tree. It is surrounded by the paler, living sapwood, which is separated from the next ring, the phloem, by a thin cambium layer. Outside the phloem is the bark-producing cork layer. Running through the rings radiating from the heartwood outward are the wood rays which are made up of parenchyma cells.

Flowering plants: Trees

Sweet chestnut *(Castanea sativa)*, far left, is a rough-barked tree with deep fissures that characteristically spiral around the trunk. The fissures occur as the cork plates in the bark pull apart when the trunk thickens. Smooth-barked trees, such as the Chinese cherry *(Prunus serrula)*, left, have a very thin bark that flakes off as the trunk thickens. The horizontal scars are lenticels, which are specialized structures that contain intercellular spaces that allow oxygen to diffuse through the bark into the trunk.

tains less moisture than sapwood. The cells often fill with bubblelike inclusions (tyloses) of waste material, such as tannins, resins, dyes, oils, gums, and mineral salts. In the teak tree *(Tectona grandis)* the inclusions are composed of silica, and in satinwood *(Chloroxylon swietenia)* of calcium oxalate.

The altered cell walls and inclusions give the timber its high degree of polish. In some species heartwood is so dense and hard that it is almost impossible to cut. Hematoxylin, which is used as a stain in the study of cells (cytology), comes from the heartwood of the logwood tree of Central and South America *(Haematoxylon campechianum)* and is a typical example of the dyes stored in trees. Tannins act as a protective antibiotic—the more tannin there is in the heartwood (making it a darker color), the more durable it is likely to be, as in, for example, mahogany *(Swietenia mahogoni)* and ebony *(Diospyros* sp.). In some species heartwood does not develop, which is why some trees, such as poplar *(Populus* sp.) and willow *(Salix* sp.), have a tendency to become hollow when they are old.

Phloem and bark

The outermost regions of the trunk, outside the cambium, are comprised of food-con-

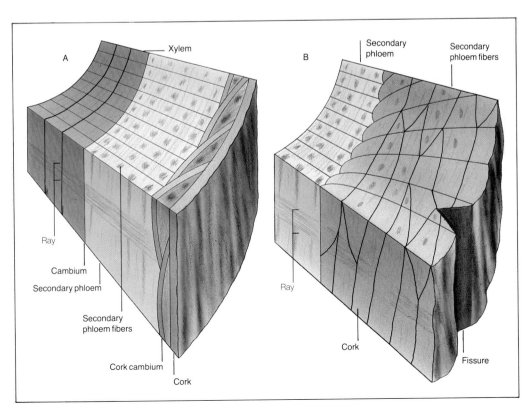

The formation of rough bark starts in the phloem when the cells differentiate to form cork cambium, which produces the cork cells (A). A second layer of cork cambium cells differentiate from parenchyma cells in the secondary phloem. Successive layers of cork cambium (B) form inside the previous layers, separated from one another by secondary phloem fibers. Each new layer cuts off the connection of the previous layer with the secondary phloem and so the outer bark dies. The cork layers tear as the trunk ages and widens, creating the deep fissures in the bark.

Flowering plants: Trees

The movement of food and water in a tree occurs in the outer layers of its branches, trunk, and roots. If a tree is stripped of these layers, or they are damaged, it will die. Water movement (marked in blue) is upward from the roots, responding to suction pressure from the transpiring leaves. Food, which is manufactured by the leaves, is usually transported downward in the phloem (indicated in red).

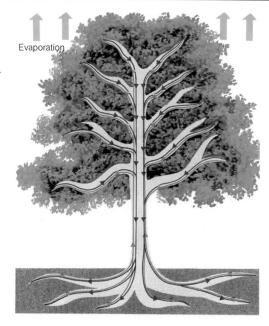

Evaporation

ducting phloem tissues. The phloem, like the xylem, consists of a number of different cell types—sieve tubes, each with a companion cell, parenchyma, and fibers. Whereas the flow of water through the xylem is upward, the movement through the phloem just beneath the bark is mainly downward, carrying foods and nutrients from the leaves down to the roots (and up again in the spring in temperate regions).

As the trunk increases in thickness, through cell division in the cambium layers, the phloem often forms wedges that taper toward the periphery (this is very obvious in oak trees). The spaces between the blocks of conducting phloem are filled with wedges of parenchyma cells: greatly widened phloem rays, continuous with the xylem rays in the wood, which function in the storage of starch and tannins. They remain active until cut off from the living part of the plant by a layer of cork cambium.

As a woody stem increases in width, a cork cambium forms in the outer layer of the phloem. On its outer surface it produces small, regular, rectangular cells whose walls become impregnated with a water-repellent material called suberin. These cells are cork cells and form a protective layer around the trunk that prevents evaporation, penetration of pests and disease organisms, and, in some species, is fire-resistant. The cork oak (Quercus suber) in Europe is unusual because after being damaged it develops a thick cork layer. This can later be stripped off for human use.

The bark of a tree consists of the living inner bark (secondary phloem) and dead outer bark. In a smooth-barked tree, the outer bark is superficial. In rough-barked trees, however, new outer bark forms deeper and eventually cuts off the outside cells from the inner tissue. The outer layer—a mixture of many tissues—then forms a thick, dead, outer bark. As the trunk expands, the bark cracks in a pattern characteristic of the species, while new bark is continually formed.

The color of bark is due largely to the presence of tannins, which also prevent decay of the bark cells. Spaced out over the bark are tiny pores (lenticels) filled with loosely-packed cells without suberin. They open at the surface and allow the cells of the trunk to breathe.

Transpiration

One of the problems a tree has to overcome is the transportation of water to the top of its crown. This is achieved by "suction pressure" from the leaves. As water is lost from the leaves by evaporation through transpiration, more water is drawn up the stem through the xylem from the roots, and the water in the conducting tissues of the xylem is under considerable tension. The leaf area in a large tree may be sufficient to evaporate many gallons of water in an hour. The amount of water lost during transpiration varies with weather con-

Deciduousness is characteristic of many trees growing in temperate regions. The brilliant reds of birch trees (Betula spp.) are the result of the disappearance from the leaves of the green pigment chlorophyll, to reveal the secondary pigments carotene, xanthophyll, and anthocyanin.

ditions; dry, windy days increase the transpiration rate, and humid, calm days reduce it. The amount of water lost from a transpiring forest may be greater than from an open water surface of the same area, such as a lake, because of the very large leaf surface.

Deciduousness

The leaves, being the main source from which water is lost by the tree, may be shed during periods of water shortage. Some trees regularly shed all their leaves in temperate regions at the beginning of winter, and in tropical regions, at the beginning of the dry season. These trees are said to be deciduous.

There is often a wintertime water shortage in temperate and boreal regions, but that is not the only reason for leaf loss. Low temperatures would destroy the living cells as ice would form in the leaves. Leaf fall is the response to day length. So deciduous trees of temperate and boreal regions shed their leaves and remain dormant throughout winter.

The mechanism that causes the leaves to fall involves processes whereby the leaf is separated from the twig on which it is growing without the twig being damaged and at the same time protecting the exposed surface from drying out and enabling infection. Two layers form—the separation and the protective layers. At the leaf base there is a structurally weak zone known as the abscission layer. After food reserves and mineral nutrients have been withdrawn into the woody branches, a periderm grows over this zone at the leaf base, and eventually the leaf falls off, leaving a scar.

Before the leaf abscises, the chlorophyll in it breaks down, and it loses its green color. Secondary pigments, such as carotene and xanthophyll, become visible and give the leaves their autumnal yellow, orange, and purple colors.

Evergreen trees retain some of their leaves all year round. Each leaf has a life of about three to four years, and there is a continual exchange of leaves. To conserve water, evergreen leaves are often covered with a thick, waxy cuticle, which gives them a glossy appearance. Deciduous leaves have a much thinner cuticle and tend to be softer and less glossy.

Roots

The crown and trunk of a tree depend on the roots for anchorage, water, and mineral salts. The vascular structure of the roots is much simpler than that of the trunk, and only the bigger roots are surrounded by bark. The roots branch laterally, crossing over each other, and may fuse to form a network. There are two types of root—thick, cablelike roots which anchor the tree, and finer, feeding roots. The anchor roots tend to grow 1 to 3 feet below the soil surface and may do so in search of water, especially in arid areas. The feeding roots mostly form a mat just below the soil surface. The finest of these have root hairs near their tips through which water and mineral salts are absorbed. In tropical rain forests the soil is typically very poor in mineral nutrients, and most of the tree roots are very near the surface to take advantage of the nutrients deposited by fallen leaves. This is also true for northern and mountain forest trees.

Flowers

Trees may have spectacular, showy flowers, such as those of the buckeye *(Aesculus octandra)* and flowering magnolia *(Magnolia grandiflora)* or they may be insignificant, like those of the ash *(Fraxinus Americana)* or oaks *(Quercus* spp.). The difference in the appearance of the flowers is due to their different methods of pollination. Large, bright flowers attract animals to

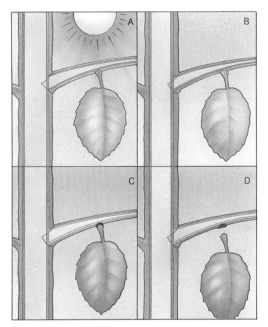

In summer (A) when days are long and warm, the leaves of trees can photosynthesize at a fast rate, drawing vast amounts of water from the roots to replenish the water loss through transpiration. As the days shorten and temperatures drop, food reserves are withdrawn from the leaves, and the chlorophyll breaks down. Secondary pigments become apparent and give the leaves their autumnal colors (B). An abscission layer grows across the base of the leaf stalk (C) and finally the leaf drops off the branch (D).

Large root systems are needed to anchor trees, which are top-heavy structures. Frequently, more than half a tree's bulk will be in the form of roots underground. These roots also serve to extract water and mineral nutrients from the soil, which are transported by the xylem to other parts of the tree.

Flowering plants: Trees

The vivid scarlet of the flowers of the flame, or flamboyant tree (Delonyx regia), attracts birds and insects. These animals aid the pollination of the flowers by collecting and transferring pollen as they move from flower to flower.

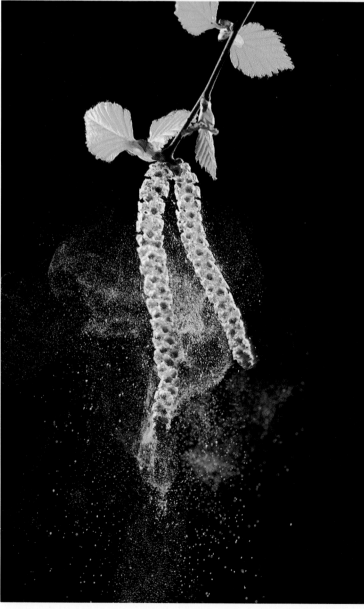

The catkins of the yellow or silver birch shed pollen as the wind shakes them, releasing from each one 5 million or so grains of pollen. Because they do not rely on animals for pollination the catkins do not have bright colors.

them. In temperate and boreal forests, these animals are insects, such as beetles, moths, and flies. In the tropics, the pollinators also include birds, such as sunbirds and hummingbirds, bats, and other mammals, such as the brush-tongued opossum of Australia. The pollinators seek out the nectar that is usually found at the base of the petals inside the flowers, pollen, or the insects that feed in the depth of the flowers. These animals collect pollen grains on them as they brush against the male anthers and transfer them to the female stigmas as they move around.

Wind-pollinated flowers are not as conspicuous as animal-pollinated ones. They are commonly grouped in long, loose clusters or catkins, and have reduced sepals and petals, so that the anthers and stigmas are exposed to the wind. It needs only a gentle rustle when the anthers are ripe for a cloud of free-floating pollen to be released. The pollen grains of these are typically small and smooth, enabling them to be blown easily, whereas those of insect-pollinated flowers are large and ornamental. The anthers in wind-pollinated flowers are frequently on the end of long filaments and are, therefore, shaken with every breeze, shedding pollen. The stigmas are often sticky or hairy to catch the pollen and are carried on the end of long styles. Like other plants, trees can be hermaphrodite (having male and female parts on the same flower), dioecious (with male and female flowers on different trees), or monoecious, the male and female flower occurring on the same tree.

Seeds and fruits

The fruits that are produced by trees also show great variety. Seeds and fruits that are wind-dispersed are quite common because they have the advantage of height from which to drop and be blown. Some trees have winged seeds, the wings of which may be single, as on the jacaranda (Jacaranda spp.), or

Flowering plants: Trees

double, as found on some maples. Elms (*Ulmus* spp.) have seeds rather like a flying saucer. Parachute seeds are also common and are found on trees as diverse as willows (*Salix* spp.) and kapok *(Ceiba pentandra)*.

Many trees produce edible fruits, which may be dry or succulent. These are adapted to be carried by animals away from the parent tree. Dry fruits, such as acorns, are often taken away by animals to be buried in the ground and stored for later consumption; there they frequently take root and grow. Succulent fruits usually have hard, resistant seeds, which often need to pass through a gut before they will germinate. The seeds are deposited with their own fertilizer as the animal (usually a bird) defecates. More rarely, tree seeds are designed to be carried by water; alders (*Alnus* spp.) have a float, as do coconuts *(Cocos nucifera),* and many others are capable of floating.

The way in which red mangrove trees shed their seeds is remarkable. The seeds of *Rhizophora mangeli,* for example, germinate on the parent tree and develop a long heavy cylindrical root. The seed then drops and spears itself into the mud, before the tide can wash it away.

Classification

Most plant classification is based mainly on the structure of the flowers. The flowers of the family Magnoliaceae are among the most primitive of the angiosperms. The trees of this family, such as the tulip tree *(Liriodendron tulipifera)* and magnolias (*Magnolia* spp.), have large solitary flowers, the sepals and petals of which are in groups of three. The family has a wide distribution in North America and Asia and a number of species are popular as garden trees.

In contrast, the members of the family Loganiaceae, for example, are considered to be some of the least primitive of the dicotyledonous trees. Whereas the petals of magnolias are separate, those of the loganias (such as *Buddleia*) are fused into a symmetrical, tubular, or bell-shaped flower. The family is distributed throughout the tropical and subtropical regions with a few temperate outliers, and includes trees of the genus *Strychnos*, which provide strychnine (from *S. nux-vomica*), curare (from the bark of *S. toxifera*), and edible fruit (from *S. spinosa*).

Many other plant families contain important tree species. The rose family (Rosaceae) comprises most of the commonly cultivated temperate fruit trees, including apple (*Malus* sp.), pear (*Pyrus* sp.), plum, peach, and almond (*Prunus* sp.) trees. In Rosaceae the flowers are usually hermaphrodite, and the sepals and petals are usually five in number. In pear and apple trees the sepals are green, and the petals are generally white or pink. After pollination, the receptacle of the flower becomes fleshy and swollen and encloses the carpels to form the edible fruit.

Whereas the loganias, magnolias, and rosaceous trees are animal-pollinated, those of the beech family (Fagaceae) are wind-pollinated. The male flowers grow in catkins, but the female flowers form a spike of up to three flowers. In both the male and the female flowers, the sepals and petals together form a perianth, which may be green-brown or colorless. The flower parts are simple and reduced because of their method of pollination, rather than their advanced stage of development. The pollen is light and copious.

The brightly colored fleshy fruits produced by some trees attract animals that the trees rely on for the dispersal of their seeds. The animals eat the seeds contained in the fruits and later excrete them away from the parent plant.

Orange trees (*Citrus* spp.) attract insects with the smell and color of their flowers, which precede the fruit. After the flowers are pollinated the flower receptacles swell and become the fleshy fruit.

Special adaptations

Like all living organisms, plants must in order to survive be able to adapt to the environmental conditions under which they live. These conditions include extremes of temperature, availability of water and nutrients, and the supply of light. Not only must plants ensure their individual survival but also that of their seeds and spores if the species is to be perpetuated. Successful adaptation is essential for the evolution of new species, because it is the "fittest" or best-adapted species that survive to pass on their genes to subsequent generations. Similarly, those plants that are best adapted to their environments become dominant.

Not all plants, however, are equally well adapted to their environments. There is a great difference in efficiency and effectiveness of adaptation among species. The grasses of the prairies and steppes, for example, have adapted to continuous attacks by grazing animals for whom they provide a staple diet. The growing point of each grass is close to the ground, with the result that the plant survives even when the foliage higher up is cropped. In contrast, the Calabrian primrose (Dendrobium biggibum) is an unsuccessful species on the verge of extinction. It is unable to reproduce fast enough to survive being eaten by animals and being picked by people.

Types of adaptation

Plants adapt to their environment in many different ways. In deserts, where high temperatures and lack of water are problems, cacti and other succulents survive by storing fluids in their swollen stems. A waxy cuticle protects the plant against high rates of transpiration. Plants that resist such dry conditions are called xerophytes. Some annual herbs in arid areas limit their entire life cycle from seed germination to seed production to the few weeks of the rainy season. These annuals include many members of the daisy family (Compositae).

Life in saline environments also requires special adaptations. Plants that live in these conditions are called halophytes. Salt-marsh plants, such as the glasswort (Salicornia fruticosa), are regularly covered by seawater; they have adapted in various ways to withstand the build-up of salts within their cells that results from the influx of water caused by osmotic pressure. In certain freshwater plants, survival is possible only if the leaves are not completely submerged. The giant Amazon waterlily (Victoria amazonica) has air-filled honeycomb structures on the underside of its leaves that cause them to float.

Other plant adaptations are specific to food and light requirements. Epiphytes are plants that use other plants as a means of support. In forests they grow on trees in order to be nearer the light, drawing nutrients in solution through their roots from sites such as hollows in tree bark. The bird's nest fern (Asplenium nidus) is an example of an epiphytic plant that

The giant Amazonian waterlily (Victoria amazonica) survives in fresh water with the aid of air-filled honeycomb structures on the underside of its leaves. These structures help the leaves, which can measure up to 6 feet (1.8 meters) across, to float, and their buoyancy is strong enough to support large birds.

Special adaptations

Club mosses, mosses, and ferns can be epiphytic in forests, living high up on the branches of other plants where they can get at the light that is scarce on the forest floor. These epiphytes obtain mineral nutrients through their roots from their hosts and organic matter that accumulates on the branches.

relies on its host rather than the soil for its nutritional requirements.

Carnivorous plants, such as sundews (*Drosera* sp.), have adapted to environments with poor nutrient status by catching and "digesting" insects rather than relying on the soil to provide nitrogen and other minerals. Parasitic plants have evolved in such a way that they take what they need directly from plants that can provide it. In doing so, however, they often sacrifice their independence. Most have no chlorophyll and, thus, cannot survive on their own by means of ordinary photosynthesis. The pink-stemmed common dodder (*Cuscuta epithymum*), for example, lives as a parasite on hops and nettles, twining itself around its host's stems and penetrating them with small projections that extract the host's nutrients.

Evolution and adaptation

The Darwinian notion of "the survival of the fittest" operates at different levels in a group of plants. Within a plant community, for example, different species and individuals of the same species have to compete for the available resources. All may succeed by means of different adaptations, or some succeed and others fail. The more severe the "selection" pressure on individuals, the more restricted is the range of possible survivors. Thus, when the selection pressure is reduced, as in a garden, many variations occur.

Selection, and therefore survival, takes three main forms. Directional selection is the process by which plants are directed or "pushed" toward an optimum set of characteristics. Once these have been achieved, by natural selection, that situation as it currently exists is maintained and is known as stabilizing selection—the second form. Any plant not reaching the optimum has a decreased chance of survival unless environmental conditions change, in which case selection again becomes directional.

The third type is known as disruptive selection in which there are two or more sets of optimal characteristics for a particular environment. This occurs where temperature or moisture levels, for example, vary in the environment. Each optimal set stabilizes as before, but the divergence in characteristics may become so great that new subspecies or even new species evolve.

Carnivorous plants feed on insects, getting from them the nitrogen and other minerals that are absent from their nutrient-deficient environments. The Venus's-flytrap (*Dionaea muscipula*) has small hairs on the inside of its leaves that when touched by an insect, induce the leaves to close. Enzymes then digest the insect.

Swamps and marshes

The terms swamp and marsh describe areas on land where the soil is waterlogged because of poor drainage or frequent inundation. These regions can be divided roughly into three categories: saltwater marsh, acid (or nutrient poor) bog, and alkaline freshwater marsh or swamp. Some of these wetlands are immense, covering thousands of acres, such as the African Sudd and the Florida Everglades. They are of great ecological importance because they form carpets of vegetation that protect the soil from being washed away, and the water retained by them speeds up decomposition, which results in a rich soil (except in the acid areas).

Obtaining oxygen

The most constant feature of wetlands is the waterlogged, airless material in which the plants' roots are lodged. Without air, the roots (and therefore the plants) cannot survive. Consequently marsh plants have become modified in various ways to obtain oxygen.

The most important modification is the enlargement of intercellular air spaces in the tissue of the submerged parts of the plants. Such tissue, called aerenchyma (adapted parenchyma), allows oxygen to pass readily through the plant from the exposed stems and leaves to the submerged parts. It also provides buoyancy in the thick mud.

Some plants in coastal swamps have developed special vertical roots, the tips or loops of which project above the surface of the water and allow air to pass into them to be conducted to the parts under water. Known as root knees, or pneumatophores, they are found in some species of mangroves, such as *Rhizophora* spp., *Avicennia* spp., and swamp cypresses (*Taxodium* spp.).

Saltwater marshes

Marine marshes are subject to regular flooding by the sea; the plants inhabiting these areas grow in zones corresponding to the frequency of flooding, the depth and salinity of water, and length of submergence. Their positions are determined by the degree to which they can cope with inundation and the salinity of the water. These plants often have to cope with a salt concentration as high as 10 per cent.

Salt-tolerant plants, known as halophytes, have difficulty in obtaining water because, as a result of the high salt concentrations in the surrounding water, any "free" water is not easily removed. There is also the problem of osmotic balance—water from the plant cells may be forced out to balance the salt concentrations inside and outside them. In addition, the large amounts of minerals present in sea water are toxic to most plants. Consequently many halophytes, such as glasswort (*Salicornia fruticosa*), have thick, leathery, fleshy leaves that store and retain large amounts of water and slow down transpiration.

The primary colonizers of the mud fringes of temperate marine marshes, such as samphire and seablite (*Suaeda* sp.), are unaffected by frequent inundations and can tolerate high salinity because they have salt glands that excrete excess salt from their shoots. Their roots stabilize the mud and allow the establishment of and replacement by other plants, particularly cord grass (*Spartina alterniflora*). This plant spreads its underground stems (stolons) thickly and further stabilizes the mud, which is subsequently colonized by other plants. This colonization is known as succession and occurs in bogs and freshwater swamps.

Many of these plants are deciduous, shedding with their leaves excess ions accumulated from the sea water. In some, such as sea lavender (*Limonium* sp.), the chances of survival of the seeds are increased because they germinate only when exposed to seawater.

A zone equivalent to that in temperate saltwater marshes occurs in tropical coastal

Insectivorous plants, such as the common sundew (*Drosera rotundifolia*), obtain the minerals they need from trapped insects, which they digest. This nutrient source is essential for the survival of these plants in the nutrient-poor acid bogs in which they live. The sticky red hairs attached to the leaves of this plant curve inward to trap insects and then secrete a digestive substance onto them.

swamps dominated by mangroves, the various species each having their preferred niches. Red mangrove *(Rhizophora mangle)* establishes seaward, sending its aerial roots down into the sediments. Black mangrove *(Avicennia nitida)* grows on exposed mudflats, sending its pneumatophores up out of the muds. White mangrove *(Concocarpus erecta)* and *Languncularia racemosa* are more common in inland brackish waters of Florida and the Caribbean islands.

Seeds of red mangrove germinate on the shrub, drop into the muds and rapidly take root. Some seeds may float on the water before they are stranded in coastal muds.

Acid bogs

Small lakes and ponds often develop into bogs as the result of peat accumulation from the shore outward. Succession occurs, the end product of which may be a woodland with scattered trees and shrubs. The soils are acid and nutrient-poor, the result of *Sphagnum* mosses.

The several species of *Sphagnum* have different water tolerances. Some are aquatic and thus build a floating thick mat over the water's edge. Other *Sphagnum* mosses prefer less wet habitats. These preferences can be accommodated because of the hummock-and-hollow nature of the bog. The hummocks often are invaded by species of heather (*Erica* spp.), bilberries and cranberries (*Vaccinium* spp.), leatherleaf (*Chamedaphnia* sp.), Labrador tea (*Ledum* spp.), and cottongrass (*Eriophorum* spp.). In addition, the surface of the bog supports other species that are adapted to soils of low nutrient levels such as insectivorous plants. These include sundew (*Drosera* spp.), butterwort (*Pinguicula* spp.), and pitcher plants (*Sarracenia* spp.). The insectivorous species are photosynthetic, but they require nutrients such as potassium, phosphorus, and nitrogen, which they get from insects. Some species flower only after digesting insects with their added nutrients.

Alkaline fens

Fens differ from bogs in having a higher nutrient content from the surrounding landscape and in not being dominated by acid-forming sphagnum mosses and sedges. Some fens or freshwater swamps are dominated by single species such as papyrus *(Cyperus papyrus)* or reed *(Phragmites* spp.) in Africa. Species of rush (*Scirpus* spp.) and sedge (*Carex* spp.) dominate fens in northern Canada and Alaska. Scattered shrubs of birch *(Betula glandulosa)*, alder *(Alnus crispa)* and the tree, larch *(Larix laricina)* are common in these more nutrient-rich habitats.

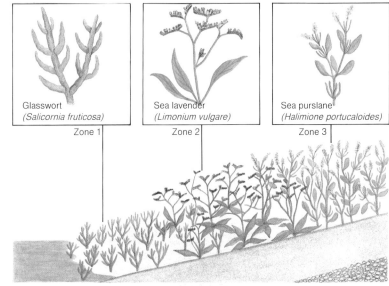

The plants on a saltwater marsh grow in zones according to their tolerance of inundation and salinity. On the seaward edges of a temperate marsh grows glasswort and other species of the genus *Salicornia*. Between the low and higher areas are a number of species including sea lavender. The drier, higher areas of the marsh are occupied by sea purslane and thrift *(Armeria maritima)*.

Mangrove swamps occur in tropical and subtropical coastal areas and include several genera, such as *Rhizophora*. In many swamps, this plant grows in the zone immediately behind the most seaward because it cannot tolerate the long periods of inundation that, for example, *Sonneratia* sp. can. These shrubs grow aerial or looped roots to trap air, which is conducted to the parts underwater. They also grow prop roots that stabilize the mud, which is then colonized by other species.

Alpine tundra

Mountains are unique habitats because they create a climate of their own. Above sea level, the mean annual temperature falls by approximately 3° or 4° F. per 1,000-foot (2° or 3° C per 300-meter) increase in elevation up to an altitude of 5 to 10 miles (8 to 16 kilometers). These zones have parallels in those that are found with increasing latitude, and the top of a high mountain in the tropics may be similar in average temperature to arctic tundra. Averages, however, can be misleading because an alpine climate shows more daily and seasonal variation than an arctic one.

The alpine tundra is the area near the top of a mountain, which is above where trees can grow (the treeline). Plants in this area show many forms of adaptation to their environmental conditions which include elevation, temperature, wind, rainfall, and snow cover.

Adaptations to temperature, wind, and snow

At very low temperatures, plant enzymes do not function efficiently and biochemical processes generally slow down. Alpine plants, such as the saxifrages (*Saxifraga* spp.), form dense cushions or tussocks that lie close to the surface of the ground, where they are protected from the wind and where the microclimate is warm. Alpine plants, such as the mountain crowfoot (*Ranunculus glacialis*), can tolerate low temperatures due to greater concentrations of sugars.

The Ruwenzori mountains dividing western Uganda and Zaire support many plants that show unusual adaptations to their climate. In the forests in the subalpine zone, temperatures alter sharply between day and night. As the temperature drops, some giant species of lobelia (*Lobelia* spp.) fold in their leaves, creating their own microclimate inside them. Temperatures of 41° F. (5° C) have been recorded inside these leafy spheres, while the temperature outside was below freezing. Rain, which is trapped in the plant's rosette during the day and warmed by the sun, also helps to keep the plant warm during the night.

Other plants, particularly on the tropical mountains, protect themselves from frost with a build-up of dead leaf material around their stems and by leaves that fold in at night to cover the growing tip. The species of groundsel (*Senecio* spp.) found in the African equatorial mountains avoid tissue damage in this way. Alpine plants are sensitive to summer heat, such as the three-leaved rush (*Juncus trifidus*), which cannot survive in conditions where the mean annual maximum temperature exceeds 72° F. (22° C). Occasionally, plants requiring higher average temperatures for growth survive on south-facing slopes but not in the valley below. The reason for this is temperature inversion, in which cold air falls to the valley floor and warm air remains above it.

In winter, strong winds in the alpine tundra, which can travel at speeds of more than 100 miles (161 kilometers) per hour and may carry lacerating ice crystals, restrict plant growth and demand special adaptations. Shrubs, such as the dwarf birch (*Betula nana*), grow close to the ground where they are less likely to be damaged by wind. Some species of willows, such as the violet willow (*Salix daphnoides*) of the Himalayas, have supple branches and stems that yield to wind pressure without breaking.

The growing season in the alpine tundra is short because snow may persist late into the summer and can be a permanent feature of north-facing hollows. Snow patches shrink gradually during the warmer months to reveal a vegetation rich in bryophytes, such as the liverwort *Anthelia juratzkana* and club mosses like the fir club moss (*Lycopodium selago*).

Snow cover can be advantageous to plants, protecting them from very low temperatures and from some wind damage. It also provides a source of water for growth early in the spring, which is necessary where soils

Mountains create unique habitats for plants and demand special adaptations if plants are to survive, reproduce, and evolve there. The great mountain buttercup (*Ranunculus lyalli,* below left) from Mount Cook and other alpine regions in New Zealand has large showy flowers to attract insects. The llareta (*Azorella glabra, below right*) lives at high altitudes in the Andes and grows as a ground-hugging cushion plant in response to low temperatures and strong winds.

Special adaptations: Alpine tundra

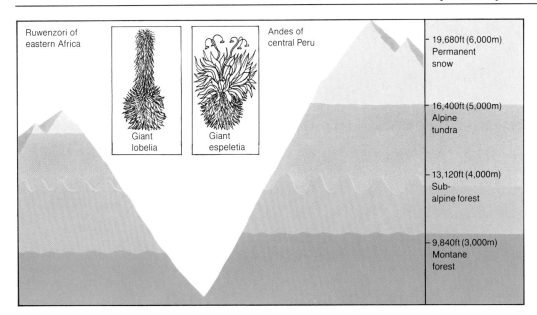

A mountain can be divided into marked climatic zones, but regional factors dictate which type of zones are present and the levels at which they lie. Vegetation in the alpine tundra is generally ground-hugging, but in the Ruwenzori Mountains, giant species of lobelia (*Lobelia* spp.) grow. In the Andes, some plants, such as the giant espeletia (*Espeletia grandiflora*), grow to a large size.

are shallow and easily drained. Many alpine plants have adaptations normally associated with warm, dry climates, such as succulent leaves, as in moss campion *(Silene acaulis)*, hairy surfaces, as in the edelweiss *(Leontopodium alpinum)*, and thick, waxy leaves, as in many of the saxifrages. Not only do these adaptations prevent the loss of water, but they also help to reflect the strong light, preventing the plant from overheating. Some species push their buds up through the melting layers to flower—for example, the alpine snowbell *(Soldanella alpina)*.

Growth and pollination

In the alpine tundra, annual plants are rare because the growing season is too short for them to complete their life cycle and produce ripe seeds. In the Alps, there are usually less than two months of the year that are frostfree. Most herbaceous plants that grow on mountains are long-lived perennials, which cope with the rigors of the climate by producing only a small amount of growth each year.

Low temperatures may result in a lack of suitable insects for pollination. There is much competition between plants for available pollinators, and flowers have developed adaptations to make themselves particularly attractive. The flowers of many alpine plants are large and showy and some, such as those of the mountain avens *(Dryas octopetala)*, take the form of a parabolic reflector that focuses the sun's rays to the central part of the flower, thereby raising its temperature and enabling the insects to work more actively.

Distribution of species

Some alpine plants have a wide global distribution, even though individual populations are isolated. The purple saxifrage *(Saxifraga oppositifolia)*, for example, is found in a complete circle around the Arctic and in the mountains of Britain, the Alps, the Hindu Kush, and the Rocky Mountains of North America. Fossil evidence has shown that this species was present in many of the intervening lowlands

Maroon Lake reflects the subalpine vegetation in the Rocky Mountains near Aspen, Colorado.

during the cold conditions of the last glaciation and has become fragmented in its distribution only during the last 10,000 years.

Mountains, particularly in tropical areas, can be regarded as climatic islands, which are centers of independent evolutionary development. The African mountains, such as the Ruwenzori, are rich in endemic species—that is, those species that are restricted in their distribution to one particular area. This area, for example, is known for its giant plants. It is likely that these sites were once in closer contact with one another, perhaps during periods of lower global temperature; subsequent isolation has, thus, allowed evolution to proceed independently, giving rise to new, endemic species. Most alpine species have limited powers of seed dispersal, so the tracts of other vegetation between mountains represent impenetrable barriers to their movement.

Arctic tundra

In the Arctic tundra, precipitation and daylight levels are high enough to stimulate vegetative growth for only three or four months of the year. During this short period many plants grow, flower, set seed, and die down.

Arctic and Antarctic tundra regions lie beyond the areas of normal tree growth, from about 60° N. and 60° S. to the poles; temperatures may be as low as −50° F. (−53° C) in the Arctic and −60° F. (−66° C) in the Antarctic. For part of the year there is continuous night with frequent high winds sweeping across the frigid landscape. During the long daylight hours of summer in the Arctic, temperatures range from an average of 35° to 55° F. (2° to 12° C). The ground below the surface is permanently frozen. There is little precipitation, at the most 6 to 10 inches (15 to 25 centimeters) in a year, including melted snow.

Low temperatures, wintry blizzards, strong winds, shallow soil and its scarce nutrients, and summer drought are reflected in the means that the plants that live in these regions have adapted to survive there. Of the plant species that live in these conditions, mosses and lichens are common, although grasses, sedges, and dwarf shrubs occur. The vegetation of the Arctic is rich compared with that of Antarctica, where temperatures are lower. In addition, most of the area in the Southern Hemisphere equivalent to that occupied by tundra in the north is ocean and the Antarctic ice cap.

General adaptations

Most tundra species grow close to the ground because upward growth is inhibited by limited snow cover and by harsh winds. The snow cover shields plants against wind abrasion by snow and soil particles. Cushion plants such as *Saxifraga oppositifolia* grow in a much warmer summer environment where plant temperatures often reach 75° to 85° F. (24° to 29° C).

Arctic shrubs grow upright where there is adequate winter snow to cover them. Along stream and river banks and below upland ridges species of willow *(Salix)* and birch *(Betula)* often reach heights of 3 to 10 feet (1 to 3 meters). Roots of shrubs and some sedges may extend down to the bottom of the active layer (top of the permafrost) at a depth of 2 to 3 feet (60 to 100 centimeters) while other species have only shallow roots 3 to 6 inches (7 to 15 centimeters) such as many heath species.

The North Alaskan woolly lousewort *(Pedicularis lanata)* creates its microclimate by trapping air between the hairs dispersed over its stem and buds. By raising the temperature around the plant in this way, the short reproductive season can be accelerated.

To protect themselves against the desiccating winds, a number of plants, such as some saxifrages (Saxifragaceae), have a leathery cuticle covering their leaves, which retards water loss through evaporation.

Flowering plants

Tundra plants are almost exclusively perennial. This is because the plants' life cycle may be interrupted by the oncoming winter and the entire process suspended until the following summer; annual plants therefore stand little chance of survival.

With a growing season limited to only eight to twelve weeks, reproduction must be rapid. The green alder *(Alnus crispa)*, for example, unfurls its catkins and leaves at the same time, rather than putting one out after the other as most other plants do.

The seeds of most tundra plants are very small; indeed, most weigh less than 1 milligram. These features not only facilitate dispersal by wind, but are also probably due to

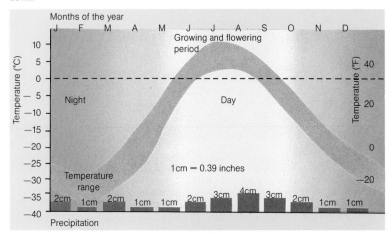

The tundra landscape is a treeless one—those shrub species that do survive are dwarfs, such as the willow *(Salix* sp.). Because of the unfavorable conditions, only a few plant species are sufficiently well adapted to this environment.

Cushion plants, such as Purple Mountain Saxifrage *(Saxifraga oppositifolia),* grow in warm summer environments where plant temperatures often reach 75° to 85° F. (24° to 29° C).

the fact that the plants need to conserve energy and cannot afford to manufacture a heavier seed. Most seeds are carried by the wind and across the snow in spring.

Some seeds lie dormant in the soil for long periods: seeds of the Arctic lupine *(Lupinus arcticus)*, for example, have been found to germinate after having been apparently frozen for thousands of years. Most seeds, however, do manage to put down some roots and grow a few leaves before the winter cold stops all growth.

Several plants also reproduce asexually, by means of rhizomes, bulbs, or root stocks, such as some species of cotton grass *(Eriophorum* spp.). This method of propagation has a far higher success rate in the Arctic conditions than seed production. These organs also serve to store nutrients.

Flowering in tundra plants is sporadic; it may be early, draining the nutrient store from the previous summer, or late, using the food manufactured in the same summer. Like the seeds, most flowers are closely packed together—there can be a hundred flowers in one square yard.

Where the land is flat and boggy, a wet tundra of grass-like sedges *(Carex* spp.) and cotton grass proliferate. Grasses also survive in these areas, growing in tussocks and thus managing to retain warmth and moisture between their leaves. The heath tundra is dominated by several species of small, compact shrubs, mostly of the berry variety, such as crowberry *(Empetrum nigrum)*, cranberry *(Vaccinium vitis-idaea)*, and bilberry *(V. uliginosum)*.

Mosses and lichens

The boggy areas where the frost has lifted the ground at intervals is called "palsa" mire. The hummocks are built up by successions of spongy bog mosses *(Sphagnum* spp.), which survive because they form low mats. Some, such as *Andreaea*, manage to live in exposed areas by growing rhizoids (little roots), which anchor the plant to the surface.

Around the fringes of the icecaps, lichens, mosses, and a few small flowering plants manage to survive. Lichens are particularly well adapted to low temperatures and conditions of prolonged drought. They are both fungi and algae combined to form a single structure, living symbiotically. The fungal structure (the outer layer) protects the plant and absorbs water vapor, while the alga (the inner layer) photosynthesizes and creates carbohydrates and other organic nutrients. Some lichens also fix nitrogen from the air. They are slow-growing, only about 1 millimeter per year.

Lichens such as reindeer moss *(Cladonia rangiferina)* are rootless and cling to other plants and rocks, especially heat-absorbing rocks in the ice-free zones. They grow when the surface they are attached to warms, and moisture is absorbed directly into the fungal cells. Growth is extremely slow because in severe weather lichens lie dormant. They reproduce vegetatively or by the dispersal of fungal spores, which join with an appropriate alga.

Woody plants in the scrub tundra, for example the dwarf willow *(Salix glauca)*, rarely grow more than 3 feet (91 centimeters) high. The small size of the plant helps to reduce its transpirational surface and, therefore, decrease water loss, as well as prevent it from being damaged by high winds.

Lava flows in time may become covered with mosses and a few flowering plants, as in Iceland.

The larch (*Larix* sp.) bears cones but, unlike other coniferous forest trees, is deciduous, losing its leaves every autumn.

Coniferous forests

In the Northern Hemisphere, north of the temperate deciduous forests, the principal plant life is coniferous forest. Because of the commercial value of the timber from these trees (used in building and as a source of wood pulp for papermaking, for example), many of the forests are man-made. The natural coniferous forest, lying mostly between latitudes 45° N. and 70° N., is called the taiga. Farther north it gives way to tundra vegetation. To the south, it blends into deciduous woodland or grassland (on the North American prairies) and steppes (in central Asia).

There are no large natural forests comparable to those of the taiga in the Southern Hemisphere because land does not extend into the equivalent latitudes, although there are some pockets of coniferous forest south of the equator in New Zealand and Chile.

Adaptations of coniferous forests

During the winter, when the ground is frozen, tree roots are unable to obtain water from the soil. Few of the winds bring rain, and although precipitation may reach 10 to 39 inches (25 to 99 centimeters) a year, most of it falls as snow.

Trees that grow in these conditions—termed winter drought—must therefore by xerophytic (drought-resistant), a feature they achieve mainly by their leaf form. The leaves of most conifers are tough, leathery, and evergreen. A waxy cuticle reduces water loss by transpiration, and the toughness prevents the leaves from wilting under water stress. Most leaves of coniferous trees are needle-shaped and highly resistant to frost.

Unlike deciduous trees, evergreen species do not need to expend as much energy putting out new green leaves each year, and those needles that persist for several years conserve scarce nutrients. Although they keep their needles, evergreens do not photosynthesize on warm winter days, for the roots remain frozen and their stomates remain closed. The trees are conical in shape, which permits light to reach lower branches for photosynthesis and permits snow to be shed so that branches do not break under the heavy weight.

Variety within the forest

Few natural coniferous forests are homogeneous, containing only evergreens. In the Siberian forest, for example, particularly at high altitudes, the dominant species are three kinds of larches, *Larix sukaczewii, L. siberian,* and *L. dahurica.* The larch is deciduous, but it is extremely hardy and can withstand gales, as well as being rot-resistant.

Parts of the Siberian forest, like large areas of Canada, occupy boggy ground called muskeg. Those in Canada support the tamarack or American larch *(Larix laricina)*, although the forests in other areas such as Labrador, New Brunswick, and Newfoundland are chiefly spruce (*Picea* sp.) and larch. White spruce *(P. glauca)* grows close in well-drained and warmer soils and black spruce *(P. marina)* in colder and wetter soils. Birch trees (*Betula* spp.) and balsam firs *(Abies balsamea)* are also found in these regions. The cells of the dormant deciduous trees remain undamaged after slow freezing. As a result, the Siberian larches can withstand lower temperatures than evergreens, and in Finland the birches extend farther north than pines and spruces.

The succession of species in a natural forest may take three centuries or more. Scattered live trees quickly grow on ground cleared by a forest fire (A), to form an open pine and spruce forest (B) that gives way about 60 years later to mixed pine and spruce (C). After a further 150 years or so, fir gradually replaces the mixed pine and spruce forest (D).

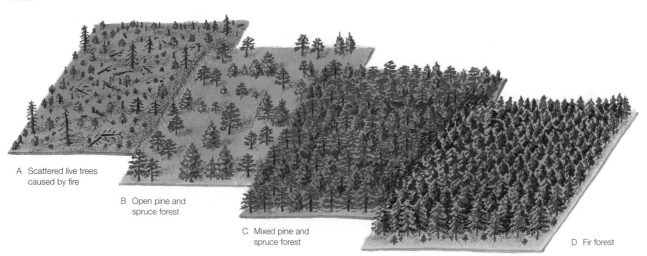

A Scattered live trees caused by fire

B Open pine and spruce forest

C Mixed pine and spruce forest

D Fir forest

The eastern side of the North American continent has a more temperate climate than western North America and Siberia. Around the Great Lakes precipitation can fluctuate between 20 and 40 inches (50 and 101 centimeters) a year and temperatures between −50° and 104° F. (10° and 40° C). These conditions result in a variety of species. White pine *(Pinus strobus)* and hemlock *(Tsuga canadensis)* dominate the forests, and red pine *(P. resinosa)* and aspen *(Populus* spp.) are common, as are black spruce *(Picea marina)* in bogs and white spruce *(P. glauca)* in uplands.

Succession of tree species

Excluding coniferous forests, there are few virgin forests left in the world. A recent study of the northern regions of Canada and Scandinavia, however, has shown how natural succession of different species—in this case of birch, pine, and spruce—occurs in one place over a period of 300 years, if it is left undisturbed.

The birch is a broad-leaved deciduous tree, which is often found in coniferous forests. When a space is cleared in a forest, by fire or falling trees, the birch (which grows rapidly and needs considerable light) quickly invades it by means of its widely dispersed, windborne seeds. For the next 60 years a birch forest is formed. Under the canopy and in gaps between the birches, pines begin to grow and the woodland changes to a pine association.

The pines, which live for about 100 to 150 years, maintain a dense forest in which ground vegetation dies because of the lack of light. Spruce and fir seedlings have difficulty in establishing themselves, but their light requirements are minimal so that, as the pine association dies away, the spruce and fir gradually become the dominant (or climax) species, so forming a spruce-fir association. Because they need less light and create great shade, the spruce and fir can maintain their dominance over the pine and birch until fire or felling renews the succession. The cycle then begins again with birch.

Where a space in the forest does admit light, juniper *(Juniperus communis)* and bilberry *(Vaccinium myrtillus)* as well as grasses and herbaceous plants, such as wintergreen *(Pyrola* sp.), grow.

Other less highly evolved plants also live in the coniferous forest. Fungi and bacteria cover old needles on the forest floor, feeding on the slowly decaying needles, and lichens such as the reindeer moss *(Cladonia rangiferina)* also grow there. Many simple plants live on the trees and are known as epiphytes. They live and grow on nutrients carried from the air by rain. They include mosses, ferns, and lichens.

The carpet of fallen needles, branches, and cones decays slowly because of the low temperatures. Beneath this layer of infertile material is a leached stratum, which the roots penetrate. It is deficient in nutrients, and so the trees depend largely on the fungi in the litter, which take carbohydrates from the tree and in turn provide it with mineral salts.

Most coniferous forests in the Northern Hemisphere are artificial plantations of fast-growing trees, raised for their timber. Often birches are left standing when the timber trees are felled.

Logs float in a collecting pond at a mill in Canada, waiting to be debarked and ground into pulp for making such materials as paper, hardboard, and insulating board. Wood pulp is also a source of cellulose for the chemical industry.

Temperate forests

Temperate forests cover large areas of the earth, occupying regions that have more than 35 inches (90 centimeters) of precipitation. Many regions experience snow and frost for four or five months in winter. The growing season is more than four months with summer temperatures averaging 65° to 80° F. (18° to 27° C) over large areas. Deciduous forests dominate the eastern United States, southeastern Canada, central Europe, and eastern China. Many other temperate forests of Australia, New Zealand, South America, and Asia have evergreen broad-leaved species.

Temperate forests can be divided into those that contain deciduous trees, and those that comprise evergreens, although the two types of tree are often found in the same forest. The regions where these forests occur can also be separated into cold, cool, and warm temperate climates, which determine the types of trees found there. Very few of the forests remain in an unmodified state—most have been felled and replanted, or selectively harvested for hundreds, if not thousands, of years.

Temperate deciduous forests

These forests usually occur in cold temperate regions, between latitudes 25° N. and 55° N., on the western fringe of Europe reaching eastward, in eastern Asia, and in the eastern United States and Canada. The Southern Hemisphere has few areas that are climatically suitable for these trees.

One of the most striking characteristics of these forests is the loss of leaves in winter and the replenishment by a new set grown in spring. During the long summer days, photosynthesis can be sustained for 16 or more hours because light levels and the availability of water are high. Deciduous trees and shrubs loose their leaves in the autumn in response to a shorter photoperiod (longer nights). The loss of leaves presents their freezing at more northern latitudes, where the roots and branches may remain frozen for several months. With cold or frozen soils, water can not be absorbed and transported to the branches. Farther south, those trees and shrubs that retain all or part of their leaf surface (live oaks) can carry on some photosynthesis in winter.

As day length becomes shorter in autumn, chlorophyll in the leaves breaks down, revealing the secondary pigments, such as carotenes, anthocyanin, and xanthophyll. These pigments enable the leaves to turn yellow, orange, and purple. The cool nights and sunny, warm days of New England, the upper Great Lakes states, and the higher mountains provide the most colorful fall foliage. Losses of important nutrients with leaf fall are minimized by the withdrawal of nutrients into the branches and roots prior to leaf fall. A thin layer of cork forms within the base of the leaf

Autumnal leaf fall occurs in response to changes in day length. A cork layer grows within the base of the leaf stem, cutting it off from the sap, and the leaf eventually drops off the branch.

In a temperate forest the dominant trees occupy the top level with smaller trees forming the shade, or second layer. Shrubs form a story of their own, below which, on the ground level, is an herbaceous layer.

stalk before leaf fall. The fallen leaves decompose to form humus, the nutrients of which are absorbed by the tree roots in the next growing season and recycled.

The wood of deciduous trees is also adapted to support the leaves in their vigorous seasonal growth—it contains vessels and tracheids that allow efficient water conduction that supports the high transpiration rate of the leaves.

A few genera of trees occur throughout these cold temperature forests, but their species vary on the different continents. They include oaks (*Quercus* spp.), beeches (*Fagus* spp.), ashes (*Fraxinus* spp.), birches (*Betula* spp.), and elms (*Ulmus* spp.). Most of the present temperate deciduous forests in Europe are dominated by beech or oak. Other trees generally found growing in association with them include maple and sycamore (*Acer* spp.), ash (*Fraxinus* spp.), and walnut (*Juglans* spp.).

Some conifers also occur in the northern deciduous forests; in the northern forests of the North American continent, for example, white pine *(Pinus strobus)* and hemlock *(Tsuga canadensis)* may be found among deciduous trees.

The cycle of leafing and leaflessness has a strong influence on the association species of deciduous woodlands. In the spring there is a period, before the canopy trees produce their leaves, when the sun can illuminate and warm the soil. At this time many of the perennial woodland herbs conduct most of their yearly photosynthesis, grow, and flower. In western Europe, species such as wood anemone *(Anemone nemorosa)*, bluebell *(Endymion non-scriptus)*, and oxlip *(Primula elatior)* flower at this time and make most of their vegetative growth. A few then produce thin, large "shade leaves," but most die down to their underground storage organs after midsummer. Many shrubs in these woodlands also flower early before the canopy trees are in leaf, such as hazel *(Corylus avellana)*.

Temperate evergreen forests

These forests are found in cool and warm temperate regions and contain both broad-leaved and coniferous species. At the lower latitudes, in the warm temperate regions, deciduous forests are replaced by evergreen, mostly broad-leaved forest. The cool temperate forests, which are coastal, are usually comprised of coniferous trees.

The evergreen broad-leaved trees, such as oaks (*Quercus* spp.) in the United States, do not need to shed their leaves because the temperatures are warm enough to keep leaves from freezing and keep soil temperatures above 39° F. (4° C); the roots can therefore constantly absorb water and sap can be supplied continually to the leaves. The leaves of these trees also have waxy surfaces and small stomata, which limit transpiration and avoid excessive water loss. Another example is the genus *Eucalyptus* in southeastern Australia. These trees also hold their leaves vertically so that they are parallel to the sun's rays and, therefore, reduce the effects of its heat.

Other warm temperate regions are also inhabited by various coniferous species, such as pines (*Pinus* spp.), as in southeastern North America, southern China, and parts of Japan. Because the humidity is so high in these areas, there is a rich understory growth of flowering herbs, mosses, ferns, and lichens.

Some cool temperate forests grow in coastal mountain slopes where sea winds bring a high annual rainfall. They also contain pines and, in North America, giant redwoods *(Sequoia sempervirens)*. Like most conifers their leaves are needlelike and waxy, which reduces transpiration, particularly in the cool winter. These trees grow densely and, like the evergreen broad-leaved forests, do not allow much light through, which results in little understory growth.

Bluebells *(Endymion non-scriptus)* carpet a birch forest for the few weeks that the warm spring sunshine can penetrate the foliage to the ground level. During this time these plants produce most of their vegetative growth. They die down to their bulbs once the leaves of the trees block out the sun, in about early summer.

Tropical rain forests

Tropical rain forests are the richest biomes on earth in terms of plant species, representing the true apex of diversity in life-forms. A single hectare may support 100 different tree species, and the microorganism and animal populations show corresponding richness. The principal forests of this kind occur in the Amazonian Basin, the Congo Basin, and from the Western Ghats of India in a belt across the Malaysian archipelago to New Guinea.

These forests vary slightly in nature depending on their latitude and altitude. In the equatorial regions temperatures range from a high of 93° F. (34° C) to a low of 68° F. (20° C), with as little range between the average temperature of the hottest and coolest months as 2° to 5° F. (1° to 3° C). Rainfall in these areas is more than 80 inches (203 centimeters) a year. Humidity, too, is high, rarely falling below 95 per cent in the lower levels of the forest. Most of these trees are evergreen. Away from the equator the rainfall drops below 78 inches (198 centimeters), and the forests in these latitudes experience a dry season, which alternates with one of copious rainfall. Many of these trees are deciduous in the drier months. The monsoon forests in Southeast Asia are also seasonal with a regular dry season with 5 inches (12 centimeters) or less of rainfall, followed by months in which the rain-bearing monsoon winds restart the seasonal growth cycle. On tropical mountains at altitudes from 2,000 to 10,000 feet (600 to 3,000 meters), a cloud or mossy forest exists, the nature of which reflects the abundant precipitation derived from fog condensing on the vegetation. Despite these differences, all tropical rain forests share a high humidity and density of vegetation, the two main features with which plants have to contend.

The forest of the Guiana Highlands marks the northern edge of the Amazonian rain forest—the biggest tropical rain forest in the world, being 2.4 million square miles (6.2 million square kilometers) in area. The density of the vegetation is clearly visible and is one of the major factors that the plants in these environments have to contend with.

Structure

The structure of tropical rain forests is complex. The tallest trees (emergents) form the so-called emergent layer, thrusting up through the forest canopy at intervals, their crowns spreading above the other trees. These umbrella crowns are usually composed of many smaller dense subcrowns. Most are between 100 and 200 feet (30 and 60 meters) tall, but some, such as the Southeast Asian tualang (Koompassia excelsa), may exceed heights of 260 feet (80 meters).

Beneath the emergent trees the canopy extends, usually as much as 150 feet (45 meters) deep. The flattened crowns of the trees interlock to form a virtually unbroken mass. Their trunks may be branchless for 65 feet (20 meters) above the jungle floor, breaking into huge spreading branches as they approach the light.

Below the main canopy are smaller trees, often with vertically elongated crowns. Some are small and slow growing, reaching maturity in the low light levels beneath the main canopy. Others are young, immature specimens of canopy or emergent trees that will mature only when they are given an opportunity to grow into spaces in those upper layers.

The understory layer beneath the enveloping canopy contains shrubs and herbaceous plants, below which—on the forest floor itself—live fungi, bacteria, algae, protozoa, and other microorganisms.

Between the layers and supported by them are other plants, such as climbers and epiphytes. Climbers are rooted in the ground, but use other plants for support as they grow toward the light. The epiphytes—orchids, bromeliads, mosses, ferns, liverworts, and lichens—do not root in the soil but live in an area from the trunk bases to the smallest twigs of the canopy, some of them at heights of 100 feet (30 meters) or more above the ground, on the branches or trunks of trees where they are exposed to the light.

The emergent layer

Towering above the body of the forest the mature emergents are subject to greater fluctuations of climate than are their companions. Winds are stronger, humidity is lower, and temperatures are higher and more extreme than beneath the shelter of the canopy. To cope with these factors, mature emergent trees develop smaller, tougher leaves than those they bear as young trees. In addition, the leaves of some species of emergents have a waxy outer covering, which helps to reduce water loss. Most of these trees are broad-leaved evergreens.

Taking advantage of the greater air movement above the canopy, many emergent species have winged fruits that are dispersed by the wind to new sites in the forest. For example, the Indo-Malayan dipterocarps (Dipterocarpaceae) have two-winged seeds, and the South American Cavanillesia platanifolia has distinctive five-winged seeds.

Special adaptations: Tropical forests

The canopy

Most canopy trees have oval, smooth, shiny leaves that taper to a point (the drip tip). A possible reason for the success of the smooth pointed leaves is that they shed rain quickly, thus discouraging the growth of tiny lichens and mosses, which flourish on moist surfaces. Even so, some species do have compound leaves, both pinnate and palmate.

In cloud forests the canopy is low and dense, formed by small trees with thick twisted crowns of tiny, leathery leaves (microphylls). Because of the intense radiation at these altitudes, as high as 10,000 feet (3,000 meters) in New Guinea, the leaves have developed a high reflective power.

Some of the canopy trees, particularly those in the drier, more seasonal areas, are deciduous, shedding their leaves at regular intervals. The Indian almond *(Terminalia catappa)*, for example, sheds its leaves every six months; other species do so at intervals of slightly more than a year.

The evergreen trees produce their new leaves in flushes rather than continuously. These new leaves tend to be more brightly colored and less rigid than the old ones. One reason for this staggered pattern of replacement may be that the soft leaves are preferred by herbivorous animals, and if these were produced continuously, the animals would destroy the bulk of new growth. By having long intervals in which no new leaves are produced, they are too uncertain a food supply to support a large population of leaf-eaters.

Flowering and fruiting occur in regular seasons. Many species, such as the silk-cotton trees *(Ceiba pentandra)*, flower simultaneously through the forest. Individuals may be widely scattered, but even so, simultaneous flowering greatly helps cross-pollination. Fruiting at the same time means that more than sufficient food for seed-eating animals and birds is pro-

Buttresses spread from the trunks of many of the canopy trees, helping to keep the trees upright and spreading their load over a larger supporting area. The roots of these trees are very shallow, their tips growing within a few inches of the soil surface, so they need this extra support.

Tropical rain forests consist of a top layer of emergent trees up to 200 feet (61 meters) tall, a dense canopy of interlocking crowns of shorter trees about 150 feet (45 meters) tall, an understory with young emergents and small, conical trees up to 50 feet (15 meters) tall, and the forest floor. On the trees are epiphytes and climbers.

duced so that enough seeds remain to germinate. In some species, flowers are produced on the main trunk rather than on twigs or branches, especially on the smaller trees. This may be related to bat pollinators, which cannot reach flowers hidden in a mass of leaves and twigs.

Most seeds in the canopy have some means of dispersal away from the vicinity of the parent tree, where competition for resources is too great. Even though the air beneath the jungle canopy is still, a few large trees have wind-dispersed seeds. The silk-cotton seeds have a light, fluffy coating that carries them on the slightest movement of air, and some of the mahoganies, *Khaya* spp. and *Entandrophragma* spp., have winged seeds.

Animals, however, are the most important means of seed dispersal in the rain forest. Attracted by soft fleshy fruits, such as durian *(Durio zibethinus)*, the animal consumes the pulp, but the hard seeds within are resistant to digestion and pass through the animal unharmed, having gained the advantage of its movement through the forest. A few trees, such as the sandbox tree *(Hura crepitans)*, have exploding fruits that scatter their seeds.

Once dispersed, the seed then has the problem of germination and establishment in the difficult environment of the forest floor, where there is tremendous competition for light and nutrients. Most plants employ one of two strategies: some trees produce a few large seeds with large food reserves, which fuel the seedling during its slow growth (this growth speeds up if a tree falls nearby, reducing the competition for light and space); other trees produce many tiny seeds, which lie on the forest floor until a gap, made by a fallen tree, allows them sufficient light to germinate.

The understory

Only 2 to 5 per cent of the sunlight available to the canopy reaches the understory, and much of the light that does remain is transmitted through or reflected off leaves, thus losing much of its useful content. When a large tree falls, possibly bringing down with it a number of smaller neighboring trees, it creates a gap where increased light levels that reach the ground layer stimulate a burst of young tree growth and the germination of seeds in the lighted patch.

Fast-growing species first take advantage of the new conditions, with fully grown canopy and emergent trees perhaps taking many decades to reestablish. The plants that do survive in the undergrowth include dwarf palms and soft-stalked species of families, such as Marantaceae (an example of which is the prayer plant, *Maranta* sp.), the ginger family (Zingiberaceae), and the acanthus family (Acanthaceae). These plants usually grow to a height of 10 feet (3 meters) or less. Their leaves are usually broad and pointed, and some species have a reddish tinge to the undersurface. The red is due to the presence of the secondary pigment anthocyanin.

Understory plants have difficulty with pollination because of the lack of air movement there. They therefore rely on insects. Some flower at night, producing large strong-scented flowers, which attract moths. Others,

The brightly colored bracts of the exotic *Heliconia rostrata* enliven the dark understory of the Amazonian rain forest. The bracts of this tall, perennial, herbaceous plant give rise to numerous flowers, which are pollinated by birds or insects.

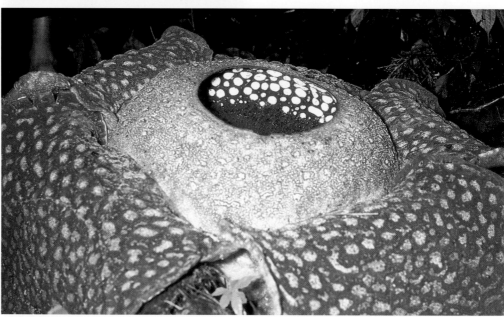

The rafflesia *(Rafflesia arnoldii)*, or monster flower, is one of the most remarkable of the flowering plants on the forest floor. Found in the Malaysian forest, it is a parasite, living inside the stems and roots of vines of the genus *Tetrastigma*. The flower that it produces above the ground is the largest known, measuring approximately 3 feet (91 centimeters) across and weighing about 5 pounds (2.2 kilograms). It smells of rotting meat, which attracts pollinating flies.

such as the cacao tree *(Theobroma cacao)* and the breadfruit plant *(Artocarpus communis)*, produce flowers and fruit on their trunks. This phenomenon, known as cauliflory, makes the flowers and large fruits conspicuous, for pollination and seed dispersal.

The forest floor

The floor itself is covered with a litter of rapidly decomposing vegetation and organisms that break it down. They are an essential component in the cycle of nutrient flow between generations of plants.

The soil of these forests is poor and intensely leached. A high proportion of the nutrients in the system are locked in the very large biomass, and there is great competition for the nutrients released by decomposition. This is one reason for the shallow-rootedness of many of the trees.

Climbers and epiphytes

Climbers attach themselves to the trunks of trees by clinging roots that may absorb water and nutrients from the surface of their supporters. The lianas, in contrast, start life as a small plant and gradually grow up to the canopy, using other plants for support. Lianas sometimes twine around other tree trunks, but often hang from the canopy, their crowns interwoven with the crowns of adjacent canopy trees. They have strange, twisted stems composed of wood that is divided into separate strands, which produce a structure of great strength and flexibility.

The epiphytes use other plants for support only. Many, particularly the epiphytic orchids, have specialized, stocky roots with a spongy cortex that takes up water quickly when it is available. Others, such as the tank bromeliads, have cup-shaped leaves or leaf bases which have been known to hold 14.5 gallons (55 liters) in some species and to support cyanobacteria and green algae.

The strangler fig *(Ficus* sp.) is one (accidental) epiphyte that does eventually kill its supporter. Its seeds are deposited by birds on the branches of canopy trees, where they germinate. Aerial roots grow down and encircle the host tree while the plant grows upward toward the light. Eventually it shadows its host, which dies from lack of light as much as from strangulation.

Tropical rain forests are a valuable resource if wisely used. Their luxuriance, however, hides a delicate ecological balance, and their complex communities are easily and irrevocably destroyed by the large-scale land clearance and logging that has taken place in recent years. The loss to humankind would be stupendous if these forests were destroyed before we even fully understand their complexity and the many species that are and will become extinct, and learn how to benefit from them.

Epiphytic bromeliads grow on the branches of an emergent tree above the western Amazonian forest, in the foothills of the Andes. These plants store water in leaf bases and survive on the nutrients of decaying bark and leaf litter in the tree hollows and stems.

A liana's woody stem is carried up toward the light by young emergent and canopy trees. Eventually its leaves will develop in the light, intermeshing with those of the canopy trees.

THE VANISHING *rain forest*

This region in the Amazon River Basin was destroyed to build a new road.

About half of the world's rain forests are gone forever—either destroyed or severely damaged—and still the destruction goes on. If the devastation continues at its present rate, some scientists predict that all the world's rain forests will be gone by 2020. These rain forests—both tropical and temperate—not only provide a home for more species than any other biome habitat, but also play a leading role in the complex global ecosystem.

The destruction of rain forests began in the 1800's in the United States, and progressed around the world through the 1900's with the growth of agribusiness, industrial forestry, and colonization. The 1960's brought a general awareness of the potential problems caused by clearing rain forests but, nevertheless, the rate of deforestation has accelerated rapidly since the 1970's.

The greatest damage is due to clear-cutting for industrialization processes, such as logging, cattle ranching, highway construction, oil drilling, dam building, and mining. But, whatever its purpose, widespread rain forest destruction is potentially disastrous to the entire planet. Although rain forests cover less than 7 percent of the earth's surface, the events that unfold in a rain forest affect life everywhere on earth. Rain forests affect all three of the pathways that make up the world's biogeochemical cycle—the water cycle, the nitrogen cycle, and the carbon-oxygen cycle. The sudden removal of its plants disrupts the rain forest's ecosystem and, consequently, the global ecosystem. Carbon dioxide levels rise, which increases the ability of the atmosphere to trap heat by admitting solar energy and preventing some surface heat from escaping out of the atmosphere. This in turn causes the global warming commonly known as the *greenhouse effect*. Due to the increase in the earth's carbon dioxide levels since the late 1800's, speculation and controversy persist about the long-term consequences of global warming. And increased carbon dioxide emissions are not the only effects of rain forest deforestation. Habitats are lost, species become extinct, rainfall increases, flooding and erosion occur, climates change, and crops fail. In addition, many rain forest species of plants and animals—still undiscovered and unstudied—are lost to us forever. Some scientists have attributed the emergence of deadly viruses such as Ebola to the destruction of their presumed habitat in the rain forests. The unknown long-term interaction of all these effects makes the final outcome of rain forest destruction difficult to predict and potentially dreadful.

Economy and ecology in harmony

Efforts to save our rain forests have taken many forms. Nations throughout the world have held conferences, signed treaties, and enacted legislation to guarantee the protection of rain forests on regional and global levels. Environmental organizations work to create awareness through public education and media attention, as well as through boycotts and demonstrations.

The most promising programs are those in which the interests of industry, environmentalists, and the people who live in rain forest regions have found a common ground. Local inhabitants of these regions, beset by both poverty and high population growth, traditionally have borne the economic burden of rain forest preservation. In the past, the pressing financial needs of daily life generally—and not unreasonably—won out over the efforts of environmentalists to preserve the forest. This situation has been a major factor in the massive rain forest destruction of the last three decades. Recently, however, environmental groups have worked to develop innovative and commercially successful ways to use rain forest products. Such efforts provide the local people with economic incentives to preserve the forests.

These new endeavors link manufacturers with local harvesters and develop markets based on the preservation of the forests rather than on their destruction. These rain forest products bring money into the local economy, give manufacturers a profit, and, through profit sharing, channel funds into future preservation efforts.

This type of cooperative effort is exemplified by the drug-discovery programs involving the pharmaceutical industries. Because rain forests are the habitat of many medicinal plants and of animals actively used in medicine, preservation of rain forest plants and animals has become a vital issue to drug manufacturers as well as environmentalists. These rain forest resources include the Pacific yew, which produces taxol, a drug used in the treatment of breast and ovarian cancer; and the rosy periwinkle, the source of vinblastine and vincristine, drugs used in the treatment of Hodgkin's disease and childhood leukemia. Vinblastine and vincristine are presently the most effective treatment for Hodgkin's disease and childhood leukemia, increasing survival rates from 2 to 58 percent and from 20 percent to 80 percent respectively.

Another cooperative effort centers around the tagua nut found in South American rain forests. Tagua nuts are now used by many manufacturers to make buttons, jewelry, and other items. The expanding line of rain forest products ranges from the Brazil nuts of the rain forest "crunch" products (similar to peanut brittle) to chocolate from cocoa beans to oils and waxes for cosmetic products. The emergence of these win-win strategies signals a new alliance between former adversaries and brings genuine hope for the future of the world's rain forests.

Seeing the forest through the trees

The long-term effects of rain forest destruction that has already occurred are the subject of many gloomy predictions, but the ultimate consequences remain unknown, and questions still outnumber answers. The challenge lies in finding the path to successful coexistence for all species in order to maintain a balance between the immediate and the long-term needs of society and the planet on which we live, and in working together as a global community of environmentally aware consumers and corporations. The journey has already proved to be hazardous, but it has begun and we are all on it.

In the Amazon River Basin in Ecuador, scientists and a native expert locate and record plants used in making medicines, (left).

Clear-cutting damage can be intensified when high winds blow down adjacent trees left unprotected, as in this section of the Olympic National Park in Washington, (below).

Heath and moorland

The terms "heath" and "moor" apply to extensive areas of land that are uncultivated and virtually treeless, with poor, acid soil—the soil tends to constitute a peat layer overlying sand or gravel. Plants of the heath family (Ericaceae) usually dominate the vegetation. They are found throughout the world usually in arctic, montane, and temperate regions with moderate rainfall and dry summers.

One of the most striking features of this type of environment is the limited number of species on such large tracts of ground. There are exceptions, however, such as the species-rich fynbos heath of the Cape in South Africa. In most areas, few species can adapt themselves to such harsh conditions, exposed to strong winds, great fluctuations in temperature, poor soil, and outbreaks of fire.

Adaptations to water stress

The winds and the sun tend to cause rapid evaporation of water from the exposed surfaces of the plants. The roots, therefore, have to absorb water from the soil to replace this loss. But the soil may be too dry to allow the water supply to be restored to its correct level. To counteract this problem, the leaves of these plants have developed various adaptations to reduce the rate of transpiration (these plants are known as xerophytes).

The leaves of heath plants are usually narrow and needlelike, as in bell heather (*Erica cinerea*), or reduced to spines, as in furze (*Ulex europaeus*). This shape reduces the transpiring surface to a minimum. The stomata, through which most water is lost, are usually confined to the lower surface of the leaves; in some species they are protected by hairs, which surround them and trap water vapor, and in others by inwardly rolled leaf margins. This rolling may be permanent, as in the heathers (*Erica* spp.), or temporary, as in many grasses, which flatten out their leaf blades when the water supply improves. These adaptations protect the stomata from drying winds and reduce water loss through transpiration.

Other water-conserving adaptations of plants in these habitats include leaves with thick cuticles, often densely crowded together and thickly covered with hairs underneath. The twigs may also be hairy.

Surprisingly, some xerophytic and xeromorphic heath plants grow in peaty, waterlogged conditions where the need for water conservation is not immediately obvious. But roots cannot survive in permanently waterlogged conditions because the soil contains little oxygen. A plant with a poor root system has restricted water uptake and, therefore, needs to control water loss.

Cushion plants and those that form mats on the ground, such as the wild azalea (*Loiseleuria procumbens*), can trap warm air among their leaves, which encourages growth. Their shape also reduces water loss and desiccation.

Nutritive adaptations

Heathland soils are poor in plant mineral nutrients, being characteristically "podsolized" (iron and humus in the upper layers are leached out by rain and carried down to the lower layers). It is thought that the sclerophyllous leaves (toughened by sclerenchyma tissue) of some xerophytic heath plants, particularly those in Mediterranean climates, may develop in nutrient-poor habitats as a response to low nutrient availability. They may also be evergreen, with a high ratio of cellulose to nitrogen. By being evergreen, they maximize the total amount of photosynthesis per unit of nitrogen over the whole life of the leaf.

Many of these plants depend on rain for a large part of their mineral nutrient requirements, but also have mycorrhizal fungi associated with their roots, such as ling (*Calluna vulgaris*) does. In this plant the fungus penetrates the cortical cells of the root and is known as an endotrophic mycorrhiza. Some plants, such as bog myrtle (*Myrica gale*), obtain nitrogen from symbiont bacteria with which they are associated; these bacteria (for example, *Rhizobium*) fix nitrogen from the atmosphere.

Plants such as common dodder (*Cuscuta epithymum*) and greater broomrape (*Orobanche rapum-genistae*) are parasites that obtain their nutrients directly from a host plant. They have no chlorophyll and so cannot manufac-

Heather (*Calluna vulgaris*), frequently dominates heaths together with members of the *Erica* genus. The tough evergreen leaves of these plants have xerophytic qualities, one of which is that the margins roll down to protect the stomata on the underside. These plants have a special relationship with little insects called thrips (order Thysanoptera) that live in the flowers, where they mature. When the winged females fly out to another flower in search of a mate (the males are wingless), they carry pollen on them that is brushed off onto the second flower, and thus cause cross-pollination. In cold climates where bees are rare, these insects are invaluable as pollinators, although heather does also rely on the wind.

ture carbohydrates. Instead they take all their requirements from their hosts. Common dodder is often found attached to the stems of heather, furze, and wild thyme (*Thymus* sp.).

Another group of heath plants, which in Europe includes eyebright *(Euphrasia officinalis)* and heath lousewort *(Pedicularis sylvatica)*, are partial parasites manufacturing their own carbohydrates, but taking mineral nutrient and water from the roots of grasses.

Fire

Because of the dryness of vegetation in summer and the open nature of the environment, fire is a frequent occurrence, but the plants have adapted to withstand its effects. The species of *Erica* are particularly resilient and grow new shoots soon after fire, from the old stem bases; the seeds, produced in large quantities, germinate freely after a fire. Those plants with underground storage organs, such as ling (with rhizomes), can die back but grow again quickly. For the same reason, bracken, furze, and purple moor grass *(Molina caerulea)* are also fire-tolerant.

The common dodder *(Cuscuta epithymum)* on heaths is frequently found attached to furze *(Ulex europaeus)* from which it feeds as a parasite. Dodder has no chlorophyll—instead, rootlike structures (haustoria) penetrate the stem of the host plant and extract mineral nutrients from it. Eventually the host plant dies.

Mediterranean-type scrub, such as chaparral (A), shares with heaths a dependence on fire. Dominant plants, such as chamise *(Adenostoma fasciculatum)*, and species of *Arctostaphylos* and *Caenothus* outlive many of the surrounding plants, which, however, have set seeds that lie dormant. These species also inhibit the growth of herbaceous plants by producing phenolic toxins, which seep into the ground around them. Fires occur naturally every 15 to 20 years (B), fueled by the dead branches of the chamise and its resinous leaves. They burn off the toxins and cut back the spreading chamise, which survives as underground stems. Fire also induces germination of dormant seeds of other plants. For a few years an herbaceous layer establishes itself (C), until the sprouting chamise starts to accumulate and exude its toxins once more.

Deserts and shrublands

Some deserts are so absolutely dry that they are almost completely devoid of plant life. The most extreme deserts, such as the Sahara in north Africa, occupy a band on about 20° north latitude. These deserts usually receive an annual precipitation of less than 4 inches (10 centimeters), and in some areas, no rain falls for years. Other deserts, such as those in the southwestern United States, northwestern Mexico, and west-central Australia, are less severe. These deserts receive an average annual rainfall of up to 10 inches (25 centimeters) all falling in one short season, which is sufficient to permit some specially adapted plants to live there.

Not only do these plants have to deal with severe drought conditions, but also the high temperatures that occur during the day—for example, up to 134° F. (57° C) in Death Valley, California—and often near-freezing temperatures at night. Temperatures are also seasonal in some deserts; those that are cold in winter are called "cold" deserts, such as the Gobi, where winter temperatures frequently drop as low as −40° F. (−40° C). Most desert plants are xerophytes (they conserve water); most also share the same means of water storage and heat endurance.

Adapted photosynthesis

Most plants have C_3 photosynthesis, but this method loses a great deal of water through transpiration. Some have adopted strategies such as C_4 photosynthesis in which carbon dioxide is fixed temporarily, released, and then refixed. The second fixing takes place in specialized vascular bundle sheath cells, and the process reduces water loss and respiration. These plants, which are mainly grasses, are found in the lower tropical latitudes. Despite their adaptation they still need some water for photosynthesis to take place. The stomata in the leaves have to open, to allow carbon dioxide to enter the plant for use in photosynthesis, which inevitably results in some water being lost. Succulents avoid this problem by keeping their stomata closed during the day, but open at night, admitting the carbon dioxide and losing less water than if they were open during the day. The carbon dioxide is then stored until daylight, when it is metabolized in a process known as crassulacean acid metabolism (CAM).

Alternatively, some plants do not photosynthesize during the hot, dry season, or at the hottest time of the day, but survive in a considerably desiccated state until the air is more humid.

Annuals

Desert plants are able to deal with the problems posed by intense heat and an irregular water supply in two ways—they endure them with the aid of morphological and physiological adaptations, or they avoid them, for example by remaining dormant.

The "avoiders" are known as ephemerals. They are mostly annuals, which survive drought in their seed form. Few of them have morphological adaptations to their environmental conditions, and most rely on their dormant seeds for survival. The seeds germinate quickly after rain; the seedlings grow rapidly, and flowering may begin very soon so that the plants can pass through their whole life cycle from seed to seed in a matter of weeks.

Virtually all annuals have mechanisms that allow germination only after large amounts of rain. These plants also tend to vary the time of germination within a single seed crop: a phenomenon known as seed polymorphism. The seeds on the parent plant of some annuals are retained even after they are ripe. When the plant receives drops of water, the bracts that hold the seeds in place open and allow only some of the seeds to fall away at each wetting. This means that their germination can be staggered, and the chances of success are correspondingly increased. In addition to varied times of germination, the seeds contain a substance called an inhibitor that stops them from germinating. This inhibitor can be washed out by water. The seeds, therefore, need at least two phases of wetting—one to cause their release, and another to remove the inhibitor and allow germination.

C_4 **photosynthesis** involves two fixings of carbon. The first, indicated in red, occurs in the middle layers (mesophyll) of the leaf, when carbon dioxide (CO_2) enters them and combines with the 3-carbon substrate phosphoenol pyruvate to form the 4-carbon compound oxaloacetic acid. This acid is converted to 4-carbon malic acid, which is then carried to the specialized vascular bundle sheath cells. Here, CO_2 is split off, and a second fixing, marked in blue, takes place. The CO_2 combines with ribulose bisphosphate, then continues into the sugar-forming reactions.

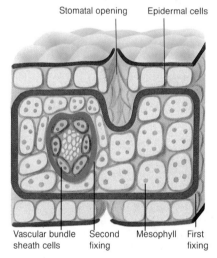

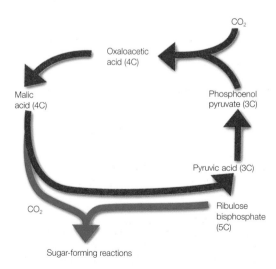

The seeds of other annuals germinate in the dark only, after a series of cycles of wetting and drying that alter the seed coat and allow free passage of oxygen to the embryo. This combination of requirements tends to cause the seeds to germinate only when buried and after several rain showers.

The strict requirements of seeds of different plants for germination cause different species of ephemerals to appear in response to rains and low or high temperatures in winter and in summer.

Annuals, like many other desert plants, depend mainly on insects for pollination. (Although the grasses and sedges are wind-pollinated.) The flowers are often large and showy and are attractive to insects, of which there may be few in these uncertain environments. The rains that stimulate plant growth, however, also encourage the emergence of insects from their dormant stages, so insects may be abundant when the plants are in flower.

Drought-deciduous plants

Another group of desert plants, which lies between the annuals and those that tolerate the desert conditions, is the drought-deciduous plants. These plants are perennials, unlike annuals, and are woody, but they avoid drought by shedding their small leaves as soon as the water availability is reduced. They remain in a state of drought-dormancy until the rain falls again, when they quickly grow a new set of leaves and flower.

Some grasses are also drought-deciduous, but they are not dormant because they rely on water and nutrients stored in rhizomes underground, until the water supply is improved.

Succulents

Succulents are also perennials. They are not woody but have a distinctive fleshy appearance because of their water-storing facility. Water is conserved in large, thin-walled cells (parenchyma tissue) in the stems or in the leaves. They usually store water in their stems, because most have reduced leaves, or none at all, such as most cacti. The thick, outer cuticle of these plants reduces dehydration, but in addition, the stomata open only at night to allow carbon dioxide entry into the plant.

A number of succulents are spherical, which is the most efficient form for water storage. The sphere is, however, a form of limited potential in a plant, and very many more desert plants have a cylindrical form, which allows taller growth.

Neither spherical nor cylindrical cacti are smooth-surfaced; almost all have longitudinal ridges or a large number of conical projections on their surface. In combination with the internal, flexible network of woody strands, the ridged surface of these plants allows expansion and contraction according to the amount of water stored in them.

Moreover, succulents possess other structural features that tend to reduce water loss. Some are pale and shiny, and reflect much of the radiation that falls on them; others are covered with white hairs that perform the same function. Some grow partially buried in the soil or sand with only the tips of the leaves

Stoneplants (*Lithops* spp.) are heavily camouflaged through their likeness to the surrounding pebbles. These succulents have no stems—the visible fleshy parts are the tips of leaves that are partly buried in the soil. Two leaves grow during each rainy season when a single flower appears between them. The old leaves then wither away.

showing on the surface. The leaf tips are transparent and are lined inside with green photosynthetic cells, which are illuminated by the light entering through the "window" in the leaf.

Desert plants are particularly vulnerable to damage by herbivorous animals, because they grow very slowly and do not recover well from damage. Many of them survive, therefore, with the help of spines, such as those found on some cacti, or detachable barbs, which discourage feeding animals; others survive by being well-camouflaged. An additional advantage for these plants is the conservation of water when they nestle among the stones.

Xeromorphs

Most xeromorphs are shrubs and trees, which tolerate drought because of the various specializations of their tough leaves. Some have small leaves (or none at all), which may be needlelike, or curled up; their small size means that less surface area is presented to the sun's heat. Xeromorphs have the stomata on one side of the leaf only (usually the upper side), which are covered when the leaf curls up; this protection has the effect of reducing transpiration, particularly in windy conditions. Others have stomata on both sides and absorb moisture from the atmosphere through them, especially when there is dew or fog around them. Those trees that do not have leaves or which lose their leaves in the dry season photosynthesize through their stems in addition to their small leaves.

Deserts support some trees, particularly around oases or watercourses where brief heavy floods recharge deep soil moisture reservoirs at long intervals. These trees have very deep roots, often reaching a depth of 100 feet (31 meters), that exploit the deep, moist soil layers.

Shrubs with xeromorphic leaves are more typical of Mediterranean-type vegetation such as California chaparral. These shrubs keep their leaves throughout the year. In some plants, such as manzanita (*Arctostaphylos* spp.), the leaves are supported vertically to reduce the area exposed to the sun, and others hold their leaves parallel to the sun's rays for the same reason. The tough structure of the leaves and their internal sclerenchymous tissue prevent the plants from wilting under water stress or from being damaged by strong winds.

Grasslands and savanna

Natural grasslands, which occur mainly in temperate regions, and savanna (tropical grassland) represent climax vegetation. The temperate grasslands of the North American prairies, the Eurasian steppes, and the South American pampas differ from those of western Europe, which are artificial. European grasslands were created through the deliberate removal of the original climax forest vegetation. They are maintained today either as meadows, for cut hay, or as pasture, for grazing farm animals.

Grasslands (except in the pampas) occur naturally in areas where the annual rainfall does not exceed 30 to 40 inches (75 to 100 centimeters) and is no less than 6 to 10 inches (15 to 25 centimeters). Seasonal differences are marked in such areas. Climate is the most important factor that defines grasslands, although several other features influence the environmental conditions, such as grazing, human activity, fire, soil structure, and topography.

Grass adaptations

One of the most significant features of grasses that enables them to survive in their habitats is their method of growth. They have closely noded underground stems that continuously produce new leaves and large numbers of shoots. Growth occurs through cell division at the bases of the leaves and stems, rather than at the tips as in most other plants. The underground stems are a means of vegetative reproduction, which allows the plant to be closely cropped above ground by grazing animals, burned by fire, die down in cold weather, or lie dormant in times of drought. In addition, the fibrous root systems and underground rhizomes of grasses and the large tap roots of numerous forbs often grow down to 6.5 feet (2 meters), where they are able to use soil water during droughts. Where precipitation is less than 10 to 20 inches (25 to 50 centimeters), roots of grasses and forbs grow down to 3 to 4 feet (90 to 120 centimeters). During periods of drought, grass leaves curl, which reduces the leaf surface, which reduces transpiration. Corn and wheat plants have the same adaptation.

Annual grasses overwinter or survive drought as seeds. These seeds germinate in late winter and early spring, utilizing soil water before the perennial native grasses begin growth. This, coupled with frequent fires and overgrazing, results in introduced annual grasses dominating what were native perennial grasslands in the western United States.

Grasses increase their chances of dispersal by having light seeds, which are carried by wind or animals. This is more important for annuals—it is their only means of reproduction.

Mountain grasslands and savannas

Many artificial European grasslands have existed since late Neolithic times. Forests were cleared, often replaced by grasses and forbs (broad-leaved, flowering plants). These converted pastures and grasslands typically occur on shallow soils overlying acid (granitic) to alkaline (limestone) rocks. These grasslands are dominated by grass species of fescue *(Festuca* spp.), bent *(Agrostis* spp.), mat grass *(Nardina stricta),* and purple moor grass *(Morlina caerulea),* along with numerous forbs. Wild species of the mint family are found in the limestone mountains of southern France and northern Italy, species that have been selected for the perfume industry.

Mountain grasslands and savannas occupy vast landscapes in New Zealand, the high Andes of South America, and smaller highlands in East Africa, Australia, Mexico, and Asia. The mountain grasslands of New Zealand are dominated by species of snow tussock *(Chinochloa* spp.) at lower elevations, with hard tussock *(Festuca novae-zelandiae),* fescue *(Festuca matthewsii)* and bluegrass *(Poa colensoi)* above. One thousand years ago, many of these lands were occupied by forests and were converted with fire by the Maoris. These grasslands, in contrast with those of South America and Australia, did not support large herbivores, so common to most natural evolved grasslands in the world. The wet

Turf-forming grasses (A) have creeping underground stems (rhizomes) with shallow, matted roots. Grasses that form bunches (B) have independent root systems and generally spread by means of their seeds. Natural meadowland (C) often forms "layers," both above and below ground. Tall herbaceous plants (forbs) bear flowers above the height of the tallest grasses; shorter forbs grow at the level of the short grasses, often blooming in spring before the tall grasses grow; and at soil level mosses and lichens thrive in the leaf litter. The various roots and root systems are similarly stratified.

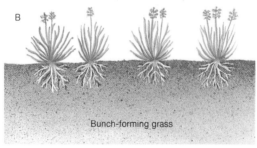

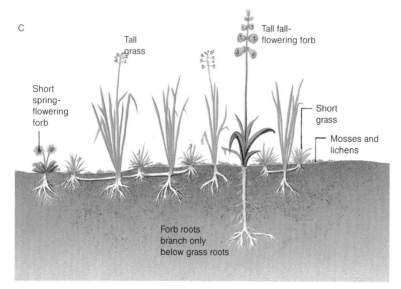

Special adaptations: Grasslands and savanna

Artificial temperate grassland on the fringes of the Black Forest includes pasture set aside for grazing animals, and meadow (foreground). Meadow grass is cut and dried to make hay, for feeding to animals in winter.

paramo of Colombia, Venezuela, and Ecuador and the dry puna of Peru, Bolivia, Chile, and Argentina are dominated by species of fescue (*Festuca* spp.), reed grass (*Calamagrostia* spp.), bluegrass (*Poa* spp.), and needle-and-thread (*Stipa* spp.). These grasses and associated forbs support the wild populations of alpaca and vicuna in the puna. The paramo and puna are used extensively for cattle- and sheep-grazing.

The Australian mountain savannas in New South Wales and Victoria, dominated by *Poa caespitosa* and *Danthonia frigida*, are grazed by sheep and cattle as well as more limited populations of kangaroo.

Temperate grasslands

Prior to the arrival of Europeans, grasslands dominated vast areas of the central and western United States and the southern portions of the prairie provinces of Canada. Tall-grass prairies extended from southern Manitoba to Texas and westward to about the 100th meridian. Most of these lands are now home to crops and pastures. The original grasslands were dominated by big bluestem (*Andropogon gerardii*), switch grass (*Panicum virgatum*), Indian grass (*Sorghastrum nutans*), and species of needle-and-thread (*Stipa* spp.). Many species of forbs, both spring- and fall-flowering, dotted these prairies, including species of the legume and sunflower families. The showy flowers of these forbs provided color to the unbroken sea of grasses.

The mixed-grass prairies that extend from southern Saskatchewan and Alberta south through the Dakotas to central Kansas, Oklahoma, and Texas are dominated by grasses and forbs, mostly less than 3 feet (1 meter) in height, where precipitation averages 20 to 25 inches (50 to 65 centimeters). These soils are not as fertile as those of the tall-grass prairie, yet they are highly productive for wheat production.

The high plains of Montana, Wyoming, Colorado to western Texas, and northern Mexico are still dominated by short-grass prairie, although large areas grow wheat with irrigation or dry-land farming. Short grasses and forbs,

The wintry steppes of Mongolia *(below left)* are natural grasslands, which support small herds of grazing ponies.

Special adaptations: Grasslands and savanna

Resembling a squadron of tanks, combine harvesters cut grain on the vast wheat fields, which have replaced the natural vegetation in Kazakhstan.

less than 15 inches (35 centimeters) in height, predominate. The sod grasses blue grama *(Bouteloua gracilis)* and buffalo grass *(Buchlae dactyloides)* and scattered forbs supported large herds of buffalo and antelope in the past and now are some of our major grazing lands. Precipitation is generally less than 15 inches (35 centimeters), often with long winters with limited snow.

Additional original grassland dominated central Washington, Oregon, southern Idaho, northern Utah, and Nevada. These lands, mostly in wheat originally, were dominated by native wheat fescue, rye, bluegrass, and needle-and-thread grasses in addition to numerous forbs. The California grasslands of the central valley were once dominated by species of needle-and-thread *(Stipa* spp.), rye grass *(Elymus)* and June grass *(Koeleria cristala)*, but now are converted to crops or to wild oats *(Avena)* and brome *(Bromus)*. The desert grasslands of southern Arizona, New Mexico, and Mexico occur in cooler and moister uplands adjacent to the Sonoran and Chihuahuan deserts. Grasses and shrubs predominate.

Other major grasslands are found in eastern Europe, Ukraine, the savannas of East Africa, the pampas of Uruguay and Argentina, the Canterbury Plains of New Zealand, and the lowland Australian savannas. In summary, many of the present and former grasslands of the world are the present world's breadbaskets in terms of crop and pasture production. Fire as well as drought played an important role in preventing tree and shrub invasion in those lands able to support woody rather than herbaceous vegetation.

Types of savanna

Many savannas occupy regions between the equatorial rain forests and the hot deserts. Because of differences in rainfall, there are three main types of savanna. The moist savanna belt that borders the equatorial forests has 42 inches (107 centimeters) or more of rain each year with a dry season lasting 5 or 6 months. Here the grasses may reach 10 feet (3 meters) in height. In the dry, savanna belt, with 24 to 42 inches (60 to 107 centimeters) of rain a year and a 7- to 8-month dry season, the grasses reach 5 feet (1.5 meters). The thorn savanna belt, with an annual rainfall of less than 24 inches (60 centimeters) and a dry season of more than 8 months, is the driest type. It has mainly annual grasses with some trees. The ground vegetation grows only to about 20 inches (50 centimeters).

Africa contains the largest single savanna region. It extends across West Africa, between the rain forests and the Sahara Desert, and sweeps through east central Africa, merging in the south into the Namib and Kalahari deserts. The African savanna includes vast tracts of each of the main savanna types.

In South America, the savanna include the llanos of Venezuela, along the Orinoco River, and the campos of Brazil, south of the Amazonian rain forest, two examples of moist savanna. Northeastern Brazil has a very dry thorn savanna with little grass.

The Australian savanna stretches across the northern part of the country. It is mostly dry savanna, some of which is difficult to differentiate from desert scrub. It is characterized by scattered Acacia trees, shrubs, and grasses.

Plant adaptations in savannas

Drought and fire, caused naturally or by farmers, have reduced the number of plant species in the savannas. The species that do occur show many adaptations to fire, and it is fire rather than climate or grazing animals that determines the stability of the vegetation.

Many trees have thick, corky, fire-resistant bark that is spongy and saturated with water after rains, as in the baobab *(Adansonia digitata)* and the bottle tree *(Cavanillesia* sp.), an adaptation that is also suited to survival

Special adaptations: Grasslands and savanna

In Australia (left), the open temperate grasslands also support scattered shrubs and trees—mostly various species of *Eucalyptus*. Sheep crop the grass among the trees.

In South American grasslands, the dominant species is often pampas grass, which produces its characteristic feathery plumes when it blooms (above). It is also grown as an ornamental plant.

during drought. Trees and shrubs also produce vast numbers of seeds, and many of the herbs have underground food storage organs.

The adaptations of grasses in temperate regions all apply equally to savanna species. One example is pampas grass (*Cortaderia* sp.) of South America, which is grown as a garden plant. In fact, gardeners are advised to burn off old leaves in spring in order to prevent the new shoots from being choked and to provide some fertilizer for the plant. Providing that the burning is rapid, the plant comes to no harm.

The soils of the tropical savanna are generally brown or black and more basic than those of the tropical forest. The dense roots of tussock-forming grasses add much organic matter to the soil.

Savanna plants

The Australian wooded savanna is dominated by gum trees—various species of *Eucalyptus* and *Acacia* spp. Like other savanna trees, they are fire-resistant. Even when the crown is severely damaged, shoots and suckers from the base and roots may ensure the tree's survival. One species of *Eucalyptus* even requires fire to allow its seeds to germinate. Fire also ensures that seedlings have a fairly free area in which to grow with a minimum of competition.

Acacias grow in both the Australian and African savannas, although they are of different species. Together with acacias in Africa are the grotesque baobabs. Elsewhere in the African savanna, from Senegal to Uganda, are such fire-resistant genera as *Terminalia* and *Isoberlinia*. *Commiphora* sp. is found with acacias in the densest savanna types, and the leguminous *Colophospermum mepane* is found in the Zambezi region in the south.

Many grasses found in the savannas are widespread species. Kangaroo grass *(Themeda trianda)* from Australia is known as red oat grass in Africa. Some tussock grasses (*Poa* spp.) occur in Australia and South America. In these areas, other typical tussock-forming savanna grasses include species of *Sporobolus, Digitaria, Panicum, Setaria, Pennisetum,* and *Sorghum* although they tend to form loose clumps rather than dense raised tussocks. Species of the genus *Hyparrhenia* (and others) have seeds that, when detached from the plant, respond to changes in humidity. Depending on whether it is wet or dry, a bristle (the awn) on the seed twists and untwists. This motion is sufficient to cause it to bury itself in the soil, where it can germinate with the next rains. This is also true of the porcupine or needle and thread grass species of *Stipa* in North America and Asia.

The African savanna (below) is the home of vast herds of wildebeest, which migrate hundreds of miles in phase with the seasonal growth of the grass. The herds are, in their turn, a source of food for predatory animals and carrion-eaters.

Aquatics

Aquatic plants (hydrophytes) survive in two main habitats—saline and freshwater. Saltwater areas include the shoreline and salt marshes; freshwater plants survive in rivers and streams, and on their banks, in lakes and in ponds. Both types are subjected to permanent immersion or frequent flooding, strong winds and water movement, and unstable soil conditions, which few other plants can tolerate. The major difference between the two kinds is that saltwater plants (halophytes) have to endure high concentrations of salt, whereas freshwater plants do not.

General adaptations

The fact that the density of completely immersed aquatic plants is similar to that of the water around them means that they have little need of support; submerged leaves and stems therefore contain little strengthening tissue.

Leaves and stems underwater also have no cuticle, so that they readily absorb dissolved carbon dioxide, oxygen, and mineral salts from the water.

Freshwater plants have a vascular system but with few woody lignified vessels in the xylem. In these plants the whole vascular system, particularly the xylem, is simplified and in some species is replaced by a cavity, or lacuna. Several aquatic species have many lacunae within their tissues. In floating leaves they are gas-filled and maintain buoyancy, raising the plant to the light so that it may photosynthesize—chloroplasts occur in the epidermal cells of the upper surface, unlike those of land plants. The lacunae also allow oxygen to diffuse rapidly from the surface to the submerged parts.

Marine habitats

Most plants that grow in the less-shallow areas of the shore are seaweeds. Those that are not tend to live in estuaries or in rock pools. Brown algae dominate in cool temperate to Arctic waters, while red algae predominate in subtropical and tropical waters.

All seaweeds require light for photosynthesis and cannot grow in water where sunlight does not penetrate. Most species have gas-filled bladders that keep them buoyant when the tide is high so that they remain near the light. All species contain chlorophyll, although red and brown seaweeds also contain other pigments (fucoxanthin and phycoerythrin respectively). Chlorophyll is not particularly efficient in trapping the bluish light that reaches deeper water, whereas the red-brown pigments (which absorb blue) gain additional energy when the plants are submerged. Brown and red seaweeds, therefore, have an advantage over green species and, in fact, only red seaweeds are found in deep water.

Apart from the algal seaweeds, there are a few flowering plants that dominate shallow marine waters, such as eelgrass (*Zostera* spp.), and species of cord grass (*Spartina* spp.) that

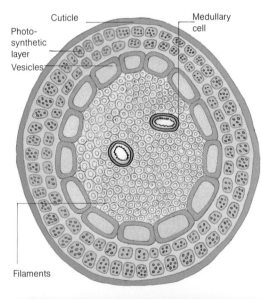

Most seaweeds (apart from the giant kelps) have no vascular system because they need no strengthening tissue, being supported by the water around them. Seaweeds have a thick cuticle that covers a thin photosynthetic layer. The inner layers are formed from densely packed, thick-walled filaments, with a few thin-walled medullary cells between them. These cells may serve to transport water through the plant.

The elongated ribbonlike leaves of aquatic plants, such as eelgrass (*Zostera* sp.), offer little resistance to moving water and are, therefore, an ideal shape in their periodically turbulent environments. This plant has submerged flowers, which are pollinated under water (one of the few flowering plants in which this occurs).

dominate the saltmarshes on coastal mud flats. Species of cord grass have vertical roots for anchorage, horizontal roots for obtaining nutrients, and stolons by which they spread, forming a compact mat within the intertidal zone.

Shore zonation

Seaweeds are often abundant on rocky shores where a clear zonation, or ecological gradient, can be seen. From the low-tide level upwards, the plants increasingly experience desiccation (drying out) and high temperatures. Many of them have developed rubbery cell walls and produce a coating of mucilage to meet these environmental stresses. The cell walls can contract and expand without damage according to the water content of the plant, and the mucilage reduces desiccation when these plants are exposed to the air.

At the lowest levels are the brown kelps (*Fucus* spp. and *Laminaria* spp.), whose long, pliable stipes (stems) make them resilient to violent wave action. Wracks dominate the middle and upper levels of the intertidal zone.

The algae that are characteristic of the lower shore often grow in rock pools at higher elevations, where they remain submerged even at low tide. These pools suffer extreme variations in temperature and salinity. Some red seaweeds, such as the thin-fronded, delicate laver bread (*Porphyra* sp.) and sea lettuce (*Ulva* spp.) that grow high on the shore, survive desiccation and recover when the tide returns.

Salt-marsh, dune, and rocky shore plants

Salt-marsh plants are subjected particularly to changing salinity levels because the salt content of the soil is increased by incoming tides and by evaporation, and then reduced by rain and dew. Halophytes, such as glasswort (*Salicornia fruticosa*), have a high osmotic concentration maintained by a high cellular concentration of amino acids (proline, for instance), which enables them to absorb water directly from the sea. Some species, such as sea lavender (*Limonium* sp.), have special salt glands to pump out excess salt onto their leaf surfaces.

Plants that grow on sand dunes and rocky shore are not true aquatics, although they may sometimes be inundated by the sea. They often have a need to conserve water, because the loose soil particles hold little water, and they are frequently buffeted by strong winds. Some have xeromorphic adaptations to reduce their transpiration rate: the prickly saltwort (*Salsola kali*) has tiny leaves reduced to spines; in some species the leaves roll up in unfavorable conditions, as in marram grass (*Ammophila arenaria*); other plants have a thick cuticle, often with a waxy coating or a thick covering of hairs. Low growth, frequently in the form of rosettes or cushions, also avoids the harmful effects of wind and evaporation. Another feature is succulence, in which leaves or stems swell when water is stored in the tissues, as in the leaves of the shrubby seablite (*Suaeda fruticosa*).

Succulence is also seen in the mangrove shrubs—so typical of the coastal swamps of tropical areas. The succulent leaves retain the water that the plant has absorbed and excrete the salt. The salt is then washed away in the frequent rains. Most mangroves have stilt roots, anchoring the plant in the mud and preventing the waves and tides from moving it. The lack of oxygen in the mud is overcome by the presence of protruding roots, or pneumatophores, which absorb oxygen from the air and take it down to the roots embedded in the mud. The zones of different mangrove species trap mud and silt, extending the land seaward. Inland from them lies a region of brackish swamps, flooded by the sea only at times of very high tide—a habitat for many tropical aquatic plants.

Rocky shores and dunes are extremely unstable habitats, where plants may be buried under sand and stones. Many plants, therefore, have extensive root systems, such as the yellow horned poppy *(Glaucium flayum)*, so that they can make new growth despite being deeply covered.

The intertidal zone of coastlines is most frequently inhabited by seaweeds, such as oarweed *(Laminaria digitata)* and thongweed *(Himanthali elongata)*. The long, flexible, but resilient, stems of these plants allow them to survive the pounding of the waves.

Special adaptations: Aquatics

Freshwater plants in fast-moving river water are usually rooted to the stones at the bottom of the riverbed. Water buttercup *(Ranunculus aquatilis)* is one of the few plants that survives this turbulence and has submerged leaves that are long and narrow and flow with the current without damage.

Common duckweed *(Lemna minor)* is a well-adapted, free-floating freshwater plant. Each plant consists of a small green thallus 3-4 millimeters across, which is kept buoyant by air-filled lacunae. It is not differentiated into stems and leaves. New cattail *(Typha)* plants are also seen here.

Freshwater habitats

Freshwater plants have to cope with bodies of water that change rapidly in their chemistry and rate of flow. The chemical composition of the water depends on the rocks over which the water has passed, the water collected from several individual springs, shallow seepage, and surface runoff. In arid regions particularly, there is a great annual fluctuation in the size and chemical properties of river water.

The variation in volume and speed of flow of river and stream water, and strong currents, can damage and uproot plants. Mountain streams are fast-flowing and, therefore, usually contain few plants; those that do survive these conditions include willow moss *(Fontinalis squamosa)*, which is anchored to stones. Sluggish rivers, however, permit plant growth in their shallows and along their margins. Still or slow-moving water in ditches, ponds, and lakes also support far more species.

Freshwater adaptations

As in deep saltwater, light does not reach the bottom of deep freshwater, and rooted plants cannot grow there. In addition, little light penetrates water that is stained with peat, mud, silt, or microorganisms. The only plants in such habitats, therefore, are free-floating species, notably algae.

Where light does reach the bottom, aquatic plants root in the mud. Some, such as stoneworts *(Chara* spp.) and Canadian pondweed *(Elodea canadensis)*, grow completely submerged. Others, such as waterlilies (Nymphaeaceae) are rooted in the bottom, but their leaves, which are attached to stalks up to 10 feet (3 meters) long, float on the surface.

Emergent aquatics grow on pond and lake edges in water less than 6 feet (1.8 meters) deep. They include reeds *(Phragmites* spp.) and bulrushes *(Scirpus* spp.), which send stems and leaves above the water surface. When these plants are the dominant species, the habitat becomes a swamp.

Freshwater hydrophytes show various adaptations to avoid damage by swamping. Many have a water-repellent cuticle on the upper surface of their floating leaves. (This also reduces water loss by evaporation.) In some species of waterlilies, the leaf margins grow vertically upward to reduce the chance of flooding. The petioles are long and, in some species, are corkscrewlike, which allows the leaves to stretch and contract to accommodate changing water levels. Diaphragms at the internodes of submerged stems prevent internal flooding if the plant is damaged.

Many aquatics have several types of leaves to cope with their watery conditions. Water buttercup *(Ranunculus aquatilis)*, for example, has divided submerged leaves and lobed floating leaves, whereas arrowheads *(Sagittaria sagittifolia)* also have aerial leaves. The floating leaves of aquatics have stomata on the upper

Special adaptations: Aquatics

surface, whereas the submerged leaves are usually long and narrow, offering minimum resistance to currents. They are often finely divided to provide a large surface for absorption. The aerial leaves are usually like those of terrestrial plants.

During the day, when freshwater hydrophytes photosynthesize, they use up the dissolved carbon dioxide, which may become scarce. Some plants, therefore, such as the quillworts (*Isoetes* spp.), have a special mechanism for absorbing carbon dioxide at night, when the respiration of aquatic animals and plants causes a carbon dioxide build-up.

The roots of freshwater plants do not take in water, but function mainly in extra nutrient absorption. They also serve as means of anchorage in bottom-rooting species and balance in free-floating species.

Most hydrophytes have flowering stems that project above the surface. The stems are usually supported by floating leaves, but such plants as waterlilies have floating flowers. Aerial flowers are pollinated by the wind or insects. A few species, however, have flowers that are adapted for pollination by water. The free-floating hornworts *(Ceratophyllum demersum)*, for example, bear tiny flowers, which open underwater. Their stamens become detached, float, and burst, releasing pollen grains that sink slowly, reaching the submerged stigmas. During floods, when aquatic plants cannot easily produce aerial flowers, submerged, cleistogamous (non-opening) and self-pollinating flowers may be produced. The seeds of hydrophytes are usually dispersed by water. Some emergent aquatics, such as the bur reed (*Sparganium* sp.), often have inflated seeds that float to the edges of lakes where they germinate.

Vegetative reproduction is also common. New fronds bud from the side of duckweed (*Lemna* sp.) and break off to form separate plants, and detached pieces of Canadian pondweed will root and grow independently. The aquatic fern, *Salvinia,* reproduces by the breaking up of old stems.

Frost damage is a winter hazard to freshwater plants. Water starworts (*Callitriche* spp.) and water soldiers *(Stratiotes aloides)* avoid it by sinking to the bottom in winter and rising again in spring. Waterlilies survive by deciduousness and by storing food in their stout rhizomes, buried in the mud. In the spring new leaves grow, and food is manufactured once more. Frogbit *(Hydrocharis morsus-ranae)* produces special winter buds (turions) on stolons that sink to the bottom when the plant dies in autumn. After remaining dormant in winter, they rise to the surface and form new plants.

The carnivorous bladderwort (*Utricularia* sp.) augments its nutrient supply by sucking insects into bladderlike traps in its submerged leaves and absorbing nitrogen from their bodies. This underwater plant sends up shoots above the surface when it blooms, so that the flowers can be pollinated by insects or the wind.

In the freshwater tape grass *(Vallisneria spiralis),* the male plants flower underwater; the buds float up to the surface where they open. The flowers on the female plants rise above the water where they are pollinated by the floating male flowers. Once fertilized, the female flower is drawn underwater where the ovule develops. The mother plant dies away and the new seed develops into a new plant.

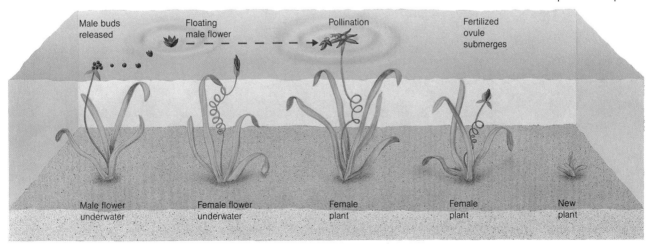

Saprophytes and symbionts

One of the most remarkable adaptations of plants to their environment is their method of obtaining food. Apart from the many autotrophic plants that manufacture food by photosynthesis there are some heterotrophic organisms that live off other organisms (as parasites), or off their decaying matter (as saprophytes) or in a mutually beneficial relationship (as symbionts). They are frequently found in inhospitable environments such as dark, dense forests, glacial polar and tundra regions, and even underground, but they also occur in our everyday environments.

Saprophytes are organisms that secure their food directly from the dead and decaying tissues of other organisms. They do so because they lack chlorophyll and cannot photosynthesize or because they contain few green parts, which allow the manufacture of only a small amount of the nutrients that they need. They include bacteria, some algae, fungi, and some flowering plants.

Symbiosis means, literally, living together. Some biological definitions restrict the meaning to a permanent association of two different organisms with a movement of metabolites between the two in which each derives an advantage from the other. Many symbioses involve an alternation between parasitism and symbiosis. Symbioses may involve an association between plants and animals, such as the algal cells in the coelenterate *Hydra,* in sea squirts and mollusks, or between different plant types, to form lichens, root-nodules, and mycorrhiza.

Saprophytes

True saprophytes secrete enzymes to break down the complex carbohydrates and proteins of food sources and then absorb the soluble foodstuffs into their cells. The most familiar saprophytes are fungi, particularly the basidiomycete mushrooms, and shelf fungi. Their hyphae form a meshed network called a mycelium, which spreads over the substrate and penetrates it. The hyphae secrete digestive enzymes that dissolve the solid components of the surrounding material; the solution is then absorbed through the membranes of the fungal body.

Most flowering saprophytes are heath species and are found in tropical Asia and Australia. They start their life as saprophytic tissue underground, nourished also by mycorrhizae; they later develop green stems and leaves. In some species, however, these parts may take several years to appear, and other species remain completely saprophytic throughout their life. In most of these species the plant lies underground except when flowering, for example Indian pipe *(Monotropa uniflora)*, but several species even flower underground, like the Australian orchids *Cryptanthemis slateri* and *Rhizanthella gardneri*.

Lichens

A lichen is a permanent association between a fungus and an alga in which the two symbionts form a single thallus. The fungus parasitizes the algal cells, extracting the carbohydrates that are formed by algal photosynthesis. It also lives saprophytically on algal cells that die. In turn, the fungus protects the alga from high light intensity and provides it with a structure that is more resistant to desiccation than the algal cell walls are.

The algal symbiont is usually a cyanobacterium (blue-green alga) or green alga; the fungal symbiont is in most cases an ascomycete, although a few lichenized basidiomycetes are known. Each species of lichen is an association of a particular algal and fungal species. Most display one of three morphologies: leafy (foliose), encrusting (crustose), and erect and tufted (fruticose).

Lichens occur worldwide and can tolerate nutrient-poor, hostile environments. They grow on exposed rock in deserts and polar regions, on solidified lava, on the bark of trees

The saprophytic orchid Indian pipe *(Monotropa uniflora)* has no chlorophyll and lives underground. It sends up shoots above ground only when it is about to flower, each shoot bearing a single bloom. The plant lives off the decaying leaf litter on the forest floor.

and on leaf surfaces (especially in tropical rain forests), on gravestones, and on the asbestos roofs of buildings. The main deterrent to lichen growth is atmospheric pollution—many species are sensitive to such pollutants as sulfur dioxide.

Root-nodules

The symbiosis between certain soil bacteria and leguminous plants to form nitrogen-fixing root-nodules is of vital importance to modern agriculture. The root hairs of legumes are invaded by the aerobic bacterium *Rhizobium* spp., which penetrates the cortical tissue and multiplies there at the expense of the host cell nutrients and enzymes. The host cells divide and enlarge to form a nodule, the cells of which become densely filled with millions of bacteria. The bacteria cells fix atmospheric nitrogen and the legume digests the bacteria, thus obtaining the nitrogen compounds for food. The nodule finally dies, and large populations of undigested bacteria return to the soil. The root-nodules of alder (*Alnus* sp.) and other trees contain symbiotic actinomycetes, which are also capable of fixing nitrogen.

Mycorrhizae

A mycorrhiza is a symbiotic association between the hyphae of certain fungi and the roots of higher plants. The mycorrhizal fungus increases the solubility of soil minerals and improves the uptake of nitrogen, potassium, and phosphorus by the host plant, protects the host roots against pathogens, and produces plant growth hormones. In return, the fungus receives a carbohydrate food supply from the photosynthetic activity of the host.

Two main types of mycorrhizae occur. In endomycorrhiza the fungal hyphae live inside the host root between and inside the cells; the fungi are usually zygomycetes, although those found in the aerial roots of tropical orchids are basidiomycetes. In ectomycorrhizae a mantle of fungal hyphae covers the root externally with some hyphae growing among the cells of the cortex; these fungi are mainly gilled or pore basidiomycetes and include *Boletus* and some puffballs, although some are ascomycetes, such as truffles (Tuberales). The ectomycorrhizae are most common on tree roots rather than roots of herbs.

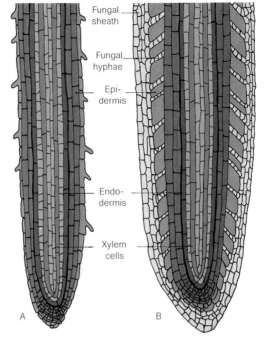

The roots of a plant without fungi (A) absorb water and what mineral nutrients they can from the surrounding soil. Many plants, however, have mycorrhizal fungi surrounding their roots (B). These fungi extract minerals more easily from the soil, which they pass to the roots through their hyphae that penetrate the root cortex. In turn, they absorb some of the carbon compounds manufactured by the plant through photosynthesis.

The encrusting lichen *Xanthoria palietina* is a common feature on gravestones.

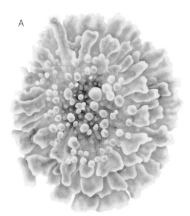
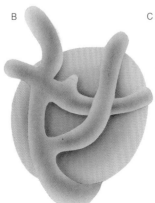
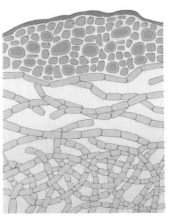

A lichen (A) forms when fungal hyphae surround an algal cell (B). Gradually the hyphae and algal cells multiply. The photosynthetic algal cells become trapped in the upper layers of the thickly intermeshed hyphae. Loosely interwoven hyphae form the middle layer of a mature lichen (C).

Plant products

People make extensive use of raw materials and chemicals derived from plants. Products such as rubber, oils, textile fibers, dyes, and pigments are present in many aspects of our daily lives. Apart from food crops, by far the most important plant resource we use is wood. It provides us with lumber and is manufactured into other products, the most significant of which is paper.

Of the 10 billion acres of land covered by forest, only about 40 per cent is exploited. The most heavily forested areas in the world include central Africa, the Amazon region, northern and southeastern Asia, and northern North America. However, Brazil, Canada, China, India, Indonesia, Nigeria, the United States, and the countries of the former Soviet Union collectively produce nearly two-thirds of the world's wood-based products. These countries have the capacity to process lumber into boards, panels, or pulp.

In many countries, the timber industry relies on conifers. Conifers are softwood and make the best pulp for paper because of the long fibers of the wood. Nearly all the trees in the forest can be utilized, and conifer forests are, thus, extremely valuable, producing more timber per acre than any other kind, including tropical forests.

Lumber

Despite the increasing use made of steel and concrete, lumber is still often employed for heavy constructions, such as bridges, harbors, and mine shafts. Some treated softwoods and some temperate and tropical hardwoods have the necessary strength, durability, and rot resistance for such building work, for example greenheart *(Nectandra rodiei)* and some members of the genus *Shorea*. Softwood plywood, which is manufactured mainly from Douglas fir and southern pine, is used extensively in the housing industry, which accounts for more than half the world's consumption of this material.

The combined physical properties of solid wood often give it an advantage over other materials. It is, for example, an excellent heat insulator. The specific heat capacity of wood is similar to metals but it is a poorer conductor of heat because it contains air in its cells. For this reason, wood is, for example, used for saucepan handles. The lightest woods, such as balsa *(Ochroma lagopus)*, are good insulators because of their large, air-filled cells and are used for containers for chemicals and liquefied gases because they keep them cool and stable.

Different types of timber may have specific qualities that can be used. One variety of willow, *Salix alba,* for example, is the only lightweight wood with the right degree of resilience to be suitable for cricket bats. Mallets and the sheaves and blocks of pulleys are made from Lignum Vitae *(Guaiacum officinale)*, a wood so dense that it sinks in water. Many musical instruments are also made from wood, but only a few high-density timbers are suitable for xylophones. Rosewood *(Dalbergia* sp.), for example, is one of the few woods to produce a musical note when struck, the length and thickness of each strip determining its pitch. Conversely, low-density materials, such as cork bark, absorb sound and are used in concert halls to reduce echoes.

Wood is one of the oldest building materials and is still used extensively today. In this concert hall, the wooden paneling has two functions: it looks attractive and also reduces echoes. Many musical instruments are made of wood. For example, spruce (*Picea* sp.) is often used for the front of violins, and bows are usually made from Brazilwood (*Caesalpinia* sp.).

Worldwide demand for pulpwood-based products, such as chipboard, fiberboard, and paper, is growing faster than the market for solid wood products. Pulp is used to satisfy the world's huge demand for paper and paperboard products.

There is a process whereby the advantages of wood can be combined with plastic. Low-quality wood is impregnated with such chemicals as vinyl acetate and exposed to gamma rays. The radiation polymerizes the vinyl compound to a hard plastic, producing a rock-hard material known as Novawood. Wood alone lacks such strength because, although the tensile strength of each cellulose fiber is stronger than steel, there are no linkages between them. It is for this reason that natural vegetable fibers, such as cotton, need to be spun and woven in order to make strong textiles.

Other plant products

The wide use of wood and other vegetable materials is dependent as much on their chemical composition as on their physical properties. Cellulose is the world's most abundant natural organic compound; it constitutes 42 per cent of wood and 95 per cent of cotton. It is a complex carbohydrate made up of about 3,000 glucose units. When chemically processed, it can be manufactured into materials, such as cellulose acetate, triacetate, and nitrate, that have a wide range of applications from plastic films to propellant explosives. Cellulose is also a beginning ingredient for making methanol and other alcohols in the chemical industry, some used as fuels.

The textile industry makes use of both raw materials and chemicals derived from plants. For example, cotton is woven in a relatively unaltered state and is used extensively for clothing and furnishing fabrics, whereas synthetic textiles, such as viscose, are chemically derived from cellulose.

The pharmaceutical industry also draws on the natural chemicals found in many plants. Aspirin was originally made from chemicals extracted from the bark of several kinds of willow (*Salix* spp.) although now the drug is artificially made.

Other plant extracts, such as oils, dyes, resins, and gums are valuable raw materials for the cosmetic, food, and printing industries. Yet another important plant extract is rubber, which is obtained by tapping the latex produced by some species of tropical and subtropical plants. More than 85 per cent of the world's natural rubber is grown in Malaysia and the Far East, the greater part of it destined for the production of tires for motor vehicles. Other uses for natural rubber are footwear, carpet backing, and conveyor belts.

Yarn is dyed in vats (foreground) using natural dyes from plants, then wound on spools (background).

Canola (*Brassica napus*) is an important oil-producing plant. It is increasingly cultivated in northern Europe, Russia, and Canada because it is a rapid grower and is resistant to cold climates. The oil is used in the preparation of margarine and in certain industrial processes.

Wood

Wood is the hard substance of trees and shrubs that makes up the trunk and the branches. It is one of the most important plant products that is used both in its natural state and in a variety of processed forms.

Two-thirds of the world's wood requirements are supplied from the boreal conifer forests, which form a zone at high northern latitudes spanning Russia, Canada, and Scandinavia. Softwoods, such as fir (*Abies* sp.), pine (*Pinus* sp.), and spruce (*Picea* sp.), grow mainly in this region, whereas hardwoods come from many parts of the world. Teak *(Tectona grandis)* and mahogany *(Swietenia mahagoni),* for example, grow in the tropical rain forests while the temperate forests are populated with many species, including oak (*Quercus* sp.), ash *(Fraxinus),* and maple *(Acer).*

Wood preparation

Before wood can be used industrially it must go through a number of processes. Felled trees have their branches removed and enter the sawmill as "green" round logs. After they have had their bark removed the logs are sawn into planks, sorted, and trimmed. The lumber must then be seasoned to match the moisture levels of its destined environment.

Wood acts like a sponge; it absorbs moisture and swells in damp air, but shrinks in dry atmospheres when the water evaporates. The wood of the balsa tree *(Ochroma lagopus),* for example, contains enough air space in its cells to absorb hundreds of times its weight in water. Seasoning not only minimizes shrinkage, but also increases resistance to fungal decay, insect attacks, and metal corrosion. In addition, seasoning makes the wood more receptive to paints, varnishes, and preservatives.

The traditional method of treating wood was air seasoning. Lumber was stacked outside under cover, each length separated by "stickers" that allowed the free flow of air. The moisture must, however, evaporate from the surface of the wood at about the same speed as it moves out from the center, or the drier surface will split and shrink. Metal cleats may be driven into dense hardwoods to prevent splitting, and sheets may be draped over very green lumber to slow surface evaporation.

Today, the faster kiln drying method of seasoning is more popular. The lumber is fed into a chamber where the temperature and humidity levels are initially low. As the moisture content of the wood falls, the humidity is reduced further, and the temperature increased until the wood reaches the desired moisture level. A fan circulates the air and dehumidifiers remove the moisture.

Fully automated kilns are costly because each type of lumber requires different treatment. Kiln drying allows for faster turnover of stock, although large planks of wood, such as oak, may still take several weeks to dry. This method of seasoning also achieves the lower moisture levels in woods that are required for precision work such as high-class joinery and flooring in modern centrally-heated houses and offices where the humidity is low.

Logs are transported by water in countries with good river systems. During the cold months of the year, the logs are piled onto the frozen rivers. When the ice melts they are floated downstream, often held together to form large rafts that are pulled to the sawmills by powerful tugboats. This is the most economical means of transportation, although large quantities of logs are also moved by rail and road.

Logs can be cut in several ways. Method A produces wide planks but these tend to warp. Methods B and C produce planks that are more stable but less economical. Waste wood can be burned to power the sawmill or processed into pulp.

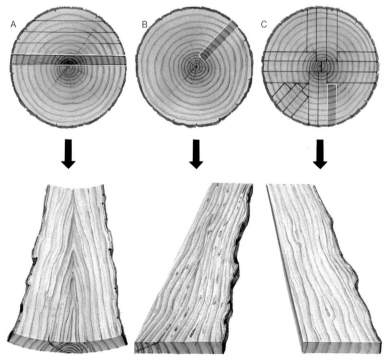

Veneer and composite woods

Many good-quality logs are sliced or peeled into thin sheets known as veneers. The logs are usually softened first to reduce the risk of splitting, and then cut to the requisite length and clamped into a lathe which turns the log against a peeling knife. The sheets are dried to a moisture content of 6 to 8 per cent and graded as face, back, or core, depending on the quality. Highly patterned veneers such as walnut (*Juglans* sp.) are used for decorative furniture. Woods that are too valuable to use in a solid form—for example, teak—are also used as veneers over a cheaper core material, such as blockboard.

Plywood was the first composite wood material to be manufactured on a large scale. It is still the most widely used today. Traditional plywood timbers include birch (*Betula* sp.) in Europe, Douglas fir (*Pseudotsuga menziesii*) in North America, and spruce in Scandinavia. Plywood is made by gluing several sheets of veneer together. The quality of each sheet in the plywood "sandwich" usually varies and the back and inner plies are generally of poorer-quality grades.

In the plywood sandwich, the grain of each veneer lies at right angles to the grain of the neighboring layer. The number of layers is usually odd, so that the grain on the two outer layers runs in the same direction. This grain-crossing structure gives the wood greater tensile strength and resistance to humidity and temperature changes because it minimizes "movement."

Because of its strength and resistance to variations in heat, plywood is used extensively for furniture and partitions. It also is used instead of solid wood when a thin but rigid material is needed. Curved plywood products, such as boats and chairs, can be made from molded plywood. The glued veneers are put between two halves of a shaped mold before being pressed. Alternatively, the plywood is molded into shape by fluid pressure inside a flexible bag.

Laminated lumber and blockboard

Laminated lumber is made from boards of seasoned lumber that are glued together with the grain running parallel in each layer but with the outer veneers at right angles to the core. For curved products, thin, malleable boards are simultaneously glued and bent to shape. Large columns, arches, boat keels, and decks are often made from this material because large planks can be made of uniformly seasoned wood and the thickness of planks can be increased at points of maximum stress. Laminated maple (*Acer* sp.), for example, is used for high-stress sports equipment, such as golf clubs, tennis rackets, and hockey sticks.

Blockboard is another form of sandwich construction consisting of a core of wooden blocks held together by surface veneers. The solid core prevents the timber from bending, and its grain runs at right angles to the veneers for strength and stability. Light woods, such as pine, may be used for the core blocks because the strength of the board lies in its construction rather than the strength of the wood.

"Sandwich" construction timbers are increasingly used in the building industry as an alternative to solid wood. The development of effective adhesives has made composite wood stronger and more adaptable.

Wood-frame construction is a traditional form of housebuilding. This method is widely employed in the United States, Canada, Scandinavia, and Australia.

Woodchip

Woodchip, made by grinding up logs or waste wood, is the raw material for making wood pulp for paper and rayon manufacture (dealt with in other articles) and the basis of what is termed reconstituted wood. The market for reconstituted wood products, such as chipboard, particle board, and fiberboard, has expanded greatly since their introduction in 1940. World demand for them is currently growing faster than that for solid wood.

The industry began as a way of recycling waste and poor quality wood, but the versatility, consistent quality, and manufacturing precision achieved by modern production processes qualify these wood-based boards for a wide range of uses, including panels for walls, ceilings, floors, doors, furniture, automobiles, and toys.

Chipboard and particle board

The production of chipboard begins when cutting blades shred shavings, splinters, and flakes of wood into tiny particles resembling coarse sawdust. These are cleaned, dried in heated tumble drums to remove moisture, and then graded. The chips are passed over a weighing belt that feeds information back to a thermal-setting resin dispenser. The correct amount of resin is dropped onto the chips in a fine spray and is mixed in by huge rotators.

The sticky chips then shower down onto a table, where they build up into mats of coarse sawdust. Thick "mattresses" form, which are first squashed in a cold press, then dampened to counteract condensation before they are clamped in a second, heated press for 10 minutes. They emerge as hard, flat chipboard. After pressing, the sheets are stacked to cool for several days to allow the chips to settle. They are then trimmed and sanded, or coated with varnish, plastic, or resin.

Extrusion is an alternative method of producing chipboard in which the chips are forced horizontally between heated metal plates that determine the thickness of the board. The resulting material has higher tensile strength, but lower bending strength.

The thickness of chipboard ranges from .25 to 1 inch (6 to 25 millimeters), with coarse chips in the middle and finer chips on the surface. A stronger, lighter board is made up of fine chips throughout. High-density board 40 to 50 pounds per cubic foot (640-800kg/m^3) is used for such products as flooring, which rely on strength and stability. Labor costs are also less for laying high-density chipboard than for laying floorboards. Medium-density board 30 to 40 pounds per cubic foot (480-640kg/m^3) is used for paneling, furniture, and shelves, whereas low-density board is best for

Woodchip is a versatile raw material that, with the addition of a thermal setting resin, can be made directly into chipboard. Alternatively it can be further broken down into wood pulp, which in turn can be made into hardboard or softboard. Wood pulp is also a source of crude cellulose for making paper, rayon, and explosives.

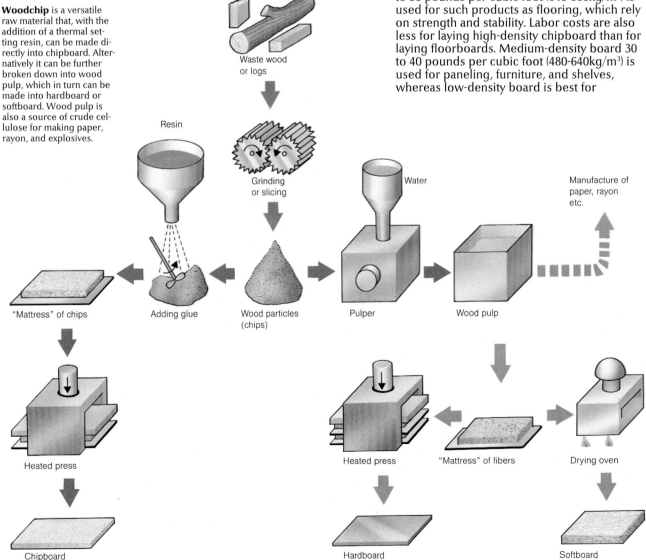

A mountain of woodchips grows beneath the conveyor at a mill (left), which has further broken down the coarse pieces in the foreground (produced by slicing logs).

A "mattress" of glue-treated woodchips (above) enters a heated press on an endless belt in a continuous process for manufacturing high-density chipboard.

insulation, ceilings, and roof deckings. All untreated chipboards are not weather-resistant.

Waferboard, a stronger version of chipboard, is made of parallel layers of long, waferlike strands of wood, which give them a resilience that matches plywood, but at one-third of the price. Other variations include hybridboard, which consists of a particle core with veneers on each surface so that it resembles plywood.

Wood-cement particle board is a product manufactured in Germany, Hungary, and Switzerland that employs cement powder instead of synthetic resins as a binding agent. As a result, the board has excellent dimensional stability and is fire-resistant. Its main potential lies in low-cost housing in tropical regions, and as wall paneling in public buildings.

Hardboard

Like chipboard, hardboard is made from forest thinnings and sawmill waste, but the chips are further reduced to fibers by steam, mincing machinery, or explosion. In the explosion process hot, high-pressure steam bombards the chips until they disintegrate. The fibers are cleaned, screened, sized, and mixed with additives, such as rosin, wax, paraffin, and chemicals, to increase their resistance to decay, fire, insects, and water.

The most popular method of producing sheets is air-felting. Resin adhesive is added to the fibers, and the mixture is fed onto a moving wire mesh to dry. In wet-felting, glue usually is not added because the softened lignin from the wood is sufficient to bind the fibers together.

The mats are cut into lengths and squeezed between rollers in a hot oven or press. The temperature and pressure determine the type and density of board.

To produce wet-felted hardboard, wet sheets are pressed in a machine with metal plates. After pressing, they may be treated with heat and humidity to improve their strength and water-resistance. Higher density boards may be tempered with oil before they are trimmed and packed to make them resistant to abrasion and weathering.

Hardboard has one smooth side and a gauze-marked reverse side. Without the gauze, which allows water and steam to escape during the pressing and heating processes, the boards would explode.

Hardboards are usually high-density, so they are suitable for flooring, shopfitting, flush doors, and furniture. For such products as kitchen cabinets, some hardboards are given a decorative overlay or a coating of enamel or plastic.

Softboard and cork

Unlike hardboard, softboard is not compressed. Instead, the wet sheets are dried in a hot-air tunnel oven to ensure that a low density of 22 pounds per cubic foot (350kg/m^3) or less is maintained. Softboard is ideal for heat and sound insulation in walls, ceilings, and roofs, where a little sagging is acceptable. Bitumen-impregnated board is used where thermal insulation combined with water-resistance is required, as in roofing.

Most cork comes from the thick outer bark of the Mediterranean cork oak (*Quercus suber*). In summer the bark is stripped away from the living trees, seasoned, and then subjected to a steaming process to soften the cells. The hard outer bark is peeled off, and the soft inner cork is pressed into sheets. Natural cork waste now supplies the raw materials for cork compositions, which are used for flooring and decorative wall tiles.

Wood pulp and paper

One-third of the world's timber is processed into wood pulp, and most of it is used to make paper and board. Pulping is necessary because woody tissue is not pure fiber, but is a rigid compound of cellulose (49 per cent), lignin (21 per cent), hemicellulose (15 per cent), and small amounts of minerals, proteins, and nitrogen. The cellulose fibers must be extracted from the binding lignin substances before they are soft enough for processing into paper and board.

In addition to forests planted specifically for pulp, one of the main sources of raw material is forest thinnings, which are trees that have been felled to allow others to grow to their full height. Wood residue from sawmills is an important alternative source of wood for pulping and has the advantage of arriving at the mill already chipped.

Waste paper is increasingly used as a raw material for both economic and ecological reasons, to the extent that the United States uses more than 30 million short tons (27 million metric tons) of recycled paper annually.

Mechanical pulping

The mechanical or groundwood method of pulping is a traditional process that involves breaking down the wood between rotating grindstones under a constant flow of water. This method produces low-grade pulp, most of which is used for newsprint. For higher quality pulp, the water pressure is increased, breaking the fibers down further. The mixture is screened, and any large lumps are removed.

The main advantages of groundwood pulping are a high yield and, therefore, a low price. This process, however, weakens the fibers so some chemical pulp is normally added to lend adequate strength. Newsprint, for example, requires about 15 per cent of chemical pulp to every 85 per cent of mechanical pulp. In addition, the residual lignin in mechanical pulp tends to go yellow with age, and so it is largely used for making "throw-away" products, such as tissues and napkins. Softwoods, such as conifers (which have long fibers and a low density), are favored for this method.

The refiner-groundwood process has developed as a result of the increased availability of wood chips from sawmills. Chips are fed into mills that reduce them to fiber fragments, and the resulting pulp, although less opaque than groundwood pulp, is much stronger; some hardwoods can be included in the mix for bulk. Nearly 50 per cent of Canada's mechanical pulp is now made by a second-generation refiner process known as "thermomechanical pulping," where the chips are preheated by steam, allowing the fibers to soften without discoloration.

Chemical pulping

Higher quality pulp can be produced by "cooking" wood chips in chemicals to remove impurities, including the lignin "glue" that binds the fibers together and causes paper to yellow. There are three main methods of chemical pulping: the sulfite or acid liquor

Before logs are processed into pulp, their bark is removed. In some factories the bark is removed by powerful jets of water. Alternatively, logs are fed into stripping drums where they jostle together until their bark peels off.

Stripped logs are cut into wood chips about one-half inch long before being pulped. In mechanical pulping, chips are mashed between grindstones under a constant flow of water, whereas chemical pulping involves "cooking" the chips in chemicals. After cleaning, the pulp is often bleached, and mixers may be added. Pulp is piped into a head box that distributes the liquid over a fast-moving mesh. Water drains out of the pulp leaving a "sheet" of fibers that are then pressed between rollers into paper.

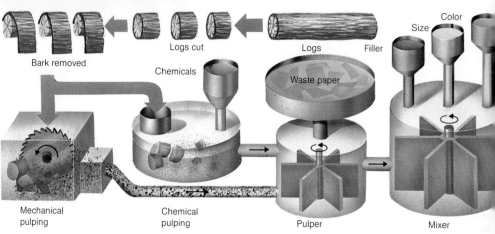

process, the soda process, and the more popular sulfate or alkaline liquor process. Sulfite liquor cooking is used mainly for spruce and some hardwoods. Inside a steam "digester," wood is pressure-cooked in a bisulfite solution mixed with sulfur dioxide gas for up to 12 hours. The pulp is then cleaned of lumps, bark remnants, and chemicals before refining. The remaining sulfite liquor, however, causes serious pollution problems if it is not disposed of carefully.

Alkaline liquor processes are used primarily for hardwoods and nonwoody fibers, such as grasses and rags. In addition they are used particularly for conifers because the alkali helps to dissolve their high resinous content. The soda process involves cooking wood chips in a caustic soda (sodium hydroxide) solution, but this method produces relatively low-grade pulp. The "kraft" or sulfate process is a more popular method where chips are cooked in a solution of caustic soda and sodium sulfide at high temperatures and under pressure. Some mills still use batch digesters in which chips are cooked in separate loads. The more efficient continuous digester allows a constant flow, which not only avoids delay but also guarantees greater uniformity in the quality of pulp.

In the digester, chips are mixed with cooking liquor, heated under pressure, and washed to remove the separated lignin and remaining liquor. The liquor-lignin mixture is burned as fuel to generate steam for the process, and the chemicals are recovered for reuse. The kraft method produces a strong, dark-brown pulp suitable for wrapping-paper.

A combination of chemical and mechanical processes is semimechanical pulp. The chips are softened in a solution of sodium sulfite and sodium carbonate or bicarbonate before they are de-fibered in a refiner.

Preparing pulp

After thorough cleansing, pulp is often bleached to transform its dirty brown color into white. Bleaching entails four stages of chemical treatment, all of which must be carefully controlled to avoid damage to the fibers. The pulp is chlorinated and then treated with caustic soda, sodium hypochlorite, and finally chlorine oxide.

To make fine printing paper, pulp must be

refined. The fibers, which are stiff and hollow at this stage, are passed through a series of metal disks that cause them to collapse, break, and fray. This process is essential if the fibers are to spread and interlock to form a strong sheet of paper. The length of time spent in the refiner determines many of the properties of the final sheet; the longer the refining process, the higher the quality of paper.

The final stage of preparation before the pulp is ready to pass to the papermaking machine is mixing, which determines the color, texture, strength, water-resistance, and opacity of the paper. China clay and calcium carbonate may be added to fill the gaps between fibers for a smoother paper and to give exceptional whiteness. Sizing agents, such as resin, are added for water-resistance, and dyes and pigments are added for color; fungicides may also be added.

Papermaking

When pulp is piped into the papermaking machine, it consists of 99 per cent water and 1 per cent fibers and additives. By the time the

The pulp-making and papermaking processes are sometimes carried out at different factories. If pulp is to be transported long distances, excess water is evaporated to produce "air dry" pulp. When this pulp arrives at the paper factory, water is added before it is fed into the papermaking machine.

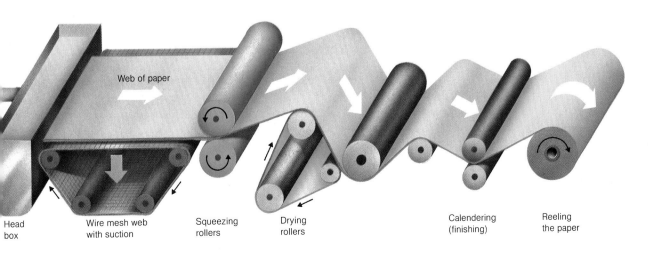

Head box | Wire mesh web with suction | Squeezing rollers | Drying rollers | Calendering (finishing) | Reeling the paper

Plant products: Wood pulp and paper

The United States is the world's leading manufacturer of paper and paperboard.

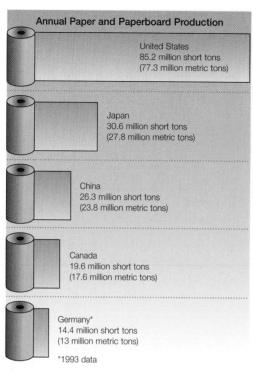

Annual Paper and Paperboard Production

United States
85.2 million short tons
(77.3 million metric tons)

Japan
30.6 million short tons
(27.8 million metric tons)

China
26.3 million short tons
(23.8 million metric tons)

Canada
19.6 million short tons
(17.6 million metric tons)

Germany*
14.4 million short tons
(13 million metric tons)

*1993 data

pulp has been processed into paper or board, it contains only 5 to 6 per cent water. Today, most papermaking machines are controlled by computers that monitor the amount of pulp flowing in, the moisture content, weight, and coating, and ensure that the finished reel of paper is consistent.

A paper machine can measure up to 300 feet (91 meters) in length and is able to produce a continuous sheet of paper more than 18 miles (29 kilometers) long in a single hour. Current trends, however, favor smaller machines that are more efficient and cost less to run, the traditional Four-drinier machine being the most widely used today.

Forming sheets

Pulp is pumped from the refiners to a pressurized head box, which ejects a continuous flow of the liquid material onto a fast-moving wire or plastic mesh. Fibers in the pulp form in the direction of the flow, and as the mesh vibrates from side to side, the fibers interlock to form a web. Much of the water in the pulp drains out, assisted by suction boxes beneath the mesh. The thickness of paper is determined by the speed of the mesh in relation to the amount of pulp released from the head box. The eventual width of the roll of the paper is dictated by the width of the mesh and can vary from 6 to 30 feet (1.8 to 9.1 meters).

Toward the end of the mesh, a cylinder covered in fine wires (a dandy roll) presses the sheet flat and an emblem may be engraved on this roll to imprint a watermark in quality bond papers. If the roll is engraved with cross-hatching, the paper will look "woven," and if parallel lines are produced, the paper is said to be "laid." The sheet is further pressed and drained of water by a suction "couch roll."

From the mesh, the sheet is transferred onto a felt blanket and a series of felted presses extract more water from it by compression. At this stage, the web of fibers has a water content of about 65 per cent before it is squeezed between the felt blanket and about 50 heated drying drums.

Some water is absorbed by the felt and some evaporates in a cloud of steam. For every ton of paper produced, about 2 tons of water evaporate. Finally, only about 5 per cent moisture remains in the paper, and this level is necessary to prevent the paper sheet from cracking.

Cardboard is made on the same principle as paper but generally by using a different technique. A cylindrical wire roll rotates in a bath of liquid pulp, and as water is drawn through the roll, a layer of fibers is deposited on the surface. The layer is transferred to a felt

Handmade paper (below left) and machine-made paper (right) are produced using the same basic technique. A watery suspension of wood pulp is floated across a wire mesh, which traps the cellulose fibers when excess water is drawn through it. The mesh is then shaken to make the fibers interlock and form a "sheet." The final size of the sheet of paper or width of the roll depends on the dimensions of the mesh.

blanket, where other layers of fiber are added to build up a thick cardboard, and then put between presses.

Coating and calendering

Coatings provide paper with a smoother, more uniform surface on which to print. They consist of such pigments as titanium dioxide or clay, mixed with adhesives and water, and are applied either by equipment in the drying process or on a separate machine. Coating may be applied to a sheet in several layers and on one or both sides of the paper until the required finish is reached.

At the end of the machine, the paper passes through a calender: a series of polished iron rollers or stacks that give the paper a glossy finish. The more rollers it passes through, the smoother the finish and the thinner the paper. A high-gloss finish is achieved by super-calendering, which involves a series of alternately stacked metal and fiber rolls, separate from the papermaking machine.

After calendering, the paper is wound into a roll around a metal core. It may be supplied as a roll or cut into sheets. High-speed sheeting machines can cut as many as eight rolls at the same time.

Other sources of pulp

Because cellulose fiber is the essential component in pulp for paper, a vast number of plants represent potential raw material for papermaking. Annual plants, for example, contain more nonfibrous cells than fibrous and produce a pulp similar to hardwood, although they demand a milder refining process. Straw is pulped in a mixture of lime and water and is still made into paper or board on a small scale in parts of Europe and Asia.

Esparto *(Stipa tenacissima)*, a Mediterranean grass, has good papermaking properties. Its leaves have a higher cellulose content than most nonwood plants, and its fibers are more uniform in size and shape. The thick-walled fibers retain their springy, sinuous form after drying and make a bulky, opaque, stable, and resilient paper.

Bagasse, the pulp obtained from sugar cane, is a useful source of paper material because it contains 65 per cent fiber, 25 per cent pith cells, and 10 per cent water-soluble substances. Provided the pith is removed during pulping, bagasse produces a relatively smooth paper and is used in Latin America and the Middle East. Bamboo has also been found to produce a satisfactory pulp and, in good conditions, yields more fiber per acre than any other plant because, although it is classified botanically as grass, its stems are unusually dense and hard like those of woody plants.

The highest quality alternative source of paper pulp is cotton rag. Rag bonds, which are usually a combination of cotton and wood fiber, are strong, fine, and smooth and suitable for bank notes, legal documents, and business letterheads. Rag papers are watermarked to specify their cotton content and are priced accordingly.

Papermaking machines are capable of producing rolls of paper more than 5 feet (1.5 meters) in diameter and more than 30 feet (9.1 meters) wide. Sometimes the rolls of paper are cut on the machine, or they may be transported whole.

The printing industry uses a large quantity of paper. Newspapers are usually made mostly from mechanical pulp, whereas high-quality magazines and books are generally made from chemical pulp.

Rayon and cellulose

Rayon is the oldest of the artificial fibers, although it is not totally synthetic (but is made by chemically modifying natural cellulose). World production averages about 2.5 million short tons (2.3 million metric tons) a year, a quantity surpassed in the textile industry only by cotton. This success is partly due to the low cost and abundance of its raw material, cellulose, which constitutes 42 per cent of wood and 95 per cent of cotton. About three-quarters of rayon output is produced from highly refined wood pulp.

Viscose rayon

Most of the world's rayon is now made by the viscose process. Sheets of wood pulp are first dissolved to form an alkali-cellulose slurry by steeping them in 18 per cent aqueous sodium hydroxide solution. After steeping, the cellulose is pressed under high pressure and as much caustic soda as possible is removed. It is then shredded into crumbs, which helps to distribute the caustic more evenly, and heated in a process called aging.

In an automatic aging room the crumbs are treated with oxygen to lower the degree of polymerization, which prevents premature hardening. Next follows the critical xanthation process, in which the mercerized alkali-cellulose is cooled and mixed with carbon disulfide to form cellulose xanthate. This yellow substance is dissolved in dilute caustic soda to form a clear syrup, known as viscose, which is filtered to remove any impurities.

Once the xanthate groups attached to the cellulose molecule have redistributed (in a "ripening" process), closer molecular chains can be formed, and the viscose is ready for spinning. First it is extruded through the holes of a spinnerette into a bath of a salt and an acid. This solution usually contains 15 per cent sodium sulfate, which removes water from the cellulose xanthate, and 10 per cent sulfuric acid, which helps to regenerate cellulose from the viscose. The extruded material forms a continuous filament containing 85-90 per cent cellulose. The size of the spinnerette determines the destiny of the yarn. A very narrow nozzle, for example, is used for making artificial hair, whereas a wide nozzle produces rayon for artificial leather fabrics. Extrusion through a narrow slit produces sheets of transparent viscose film, known as cellophane.

The fibers are spun continuously, reeled, then washed in water and bleached in sodium hypochlorite solution. After the fibers are cut to size, an appropriate finish may be applied to counteract any natural slipperiness or stickiness during weaving, before they are wound on to bobbins for shipping.

Viscose rayon fabrics—once marketed as "artificial silk"—are comfortable to wear, are cool, like cotton, but more absorbent, and can be brushed to make them warmer. Although viscose rayon creases and burns easily, it can be treated with a crease-resistant finish.

Cuprammonium rayon

Today, fibers made by the cuprammonium process, an expensive technique, are manufactured only for highly specialized purposes. Almost all artificial kidney machines, for example, use membranes prepared from cuprammonium rayon films and fibers. They are more supple, cause less bloodclotting, and have better dewatering properties than viscose rayon membranes.

In the cuprammonium process, wood pulp is mixed with aqueous ammonia, copper sulfate, and caustic soda until it produces a 10 per cent cellulose blue solution. Although it is wet spun, unlike viscose, coagulation is slow. The filaments are extruded into water, which also passes through the spinnerette, causing the fibers to elongate. As a result of this "stretch spinning," the filament size is smaller than viscose, even though the spinnerette holes are larger. The threads are hardened as they pass through a dilute sulfuric acid solution, which also removes the ammonia and copper for recycling.

Acetate rayon

The first important application of cellulose acetate was as a nonflammable varnish for fabric-covered aircraft in World War I. The first acetate yarn (known as Celanese) was developed from the redundant stocks of acetate after the war. The cellulose is steeped in acetic acid, then treated with acetate anhydride and

Viscose (right) is a sticky substance made by treating cellulose with various chemicals (see illustration on opposite page). To make rayon, the viscose is forced through tiny holes into a bath of acid, and the filaments combined and pulled from the bath as a yarn (below).

sulfuric acid. Each glucose unit in cellulose combines with three molecules of acetic acid and forms cellulose triacetate; but because it does not readily accept dyes in this form, one acetate molecule is removed (by adding water) to produce cellulose diacetate. The fibers are washed, dried, dissolved in acetone, and then extruded through a spinnerette sited 20 feet (6 meters) above the ground. As the solution descends, it is warmed by a current of hot air, which evaporates the acetone. The acetate fibers are then lubricated before twisting and winding.

The resulting cloth is more like natural silk than any other fiber. It is crease-resistant, mothproof, and water-repellent. Triacetate, sold as Tricel and now often combined with cotton, is popular for its drip-dry qualities.

The early commercial uses of cellulose acetate as photographic film, in shatterproof glass, and for contact lenses and varnishes are now threatened by acrylics, nylons, and other totally synthetic materials made from petroleum-based products. However, some experts speculate that a petroleum shortage and rising oil prices may lead to greater use of cellulose.

Nitrocellulose

The explosive properties of cellulose nitrate (often known as nitrocellulose or guncotton) are largely derived from the polymeric and fibrous structure of cellulose. It is used as a propellant explosive for rockets and guns and in cartridges for small arms. Its properties depend, however, on the extent of nitration, a nitrogen content of more than 13 per cent being necessary to make an explosive. When the nitrogen content is limited to 11 or 12 per cent, it may be used for lacquers and films, and at 10 per cent nitrogen, plastics may be manufactured from it. Low-viscosity nitrocellulose lacquers may also be produced by heating cellulose and nitrogen with water under pressure but, like other forms of cellulose nitrate, they suffer from the disadvantage of extremely high flammability.

Cellophane is a transparent material mainly employed for wrapping, especially for foodstuffs such as candy. It is chemically identical to rayon, but is produced by extruding viscose through a narrow slit so that it takes the form of a thin film or sheet.

In making rayon (below), sheets of wood pulp are steeped in alkali (sodium hydroxide, NaOH), pressed damp dry and crumbled, and, after an aging process, dissolved in carbon disulfide (CS_2) to form cellulose xanthate. This substance is reacted with more alkali to make viscose, which is filtered before being extruded into a bath containing acid. The resulting rayon yarn is washed, bleached, and dried before being wound on reels.

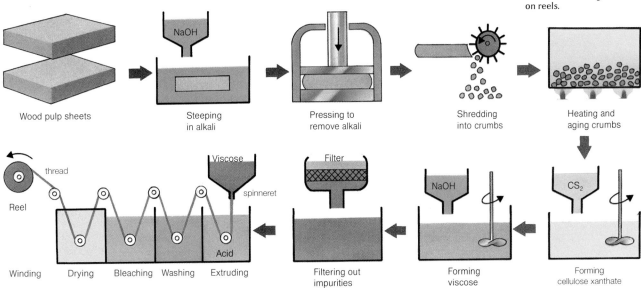

Vegetable fibers

The most important plant products after foodstuffs and timber are fiber crops. They exert economic, social, and political influence both at the local level and on an international scale. Throughout the world millions of people are employed in growing them, and major industries are based on processing the fibers into yarns and textile fabrics.

Synthetic fibers have made significant inroads into the textile market in recent years but—particularly since the increase in the price of petroleum, the base material for many synthetic fibers—the natural products remain competitive. Genetic modifications to the fiber plants have improved yields. And compared with their chief "natural" competitors—fibers made from wood pulp derived from timber—most vegetable fiber crops mature in months rather than the decades required by trees.

There are three main types of vegetable fibers, classified as soft, hard, or short. The soft (or bast) fibers are extracted from the stems of plants and include jute, flax (linen), hemp, ramie, and kenaf. Hard fibers, such as manila hemp, sisal, henequen, and New Zealand flax, are gathered from the leaves for providing hard cordage or brush and have suffered more from the impact of synthetic fibers (which can be made stronger). Cotton, the world's most important vegetable fiber, belongs to the third group, the short fibers, produced from the "hair" on the seeds; they also include kapok and coir (from coconut).

Fiber structure and characteristics

The bast and hard fiber strands are bundles of numerous overlapping, parallel-fiber cells. The cell walls consist of minute microfibrils (the smallest units visible under a microscope), which are composed of cellulose chains linked by hydrogen bonds. The orientation of the microfibrils determines each fiber's elasticity.

Cotton is unique among the natural fibers. Each fiber is a single tubelike cell made up of 20-30 layers of cellulose. When the cotton boll (fruit capsule) opens, the fibers dry into twisted ribbons with spiraling microfibrils.

A fiber must be at least 100 times longer than it is wide to be suitable for spinning into thread for weaving. Although its length measures up to 3,000 times greater than its width, cotton is relatively short—only one-third of the length of flax—and can pose problems for the spinner and weaver. Fibers of the finest Sea Island cotton *(Gossypium barbadense)* are up to 1.5 inches (38 millimeters) longer than those of coarse Indian blanket cotton *(G. herbaceum)*. Natural twist in the fibers also contributes to quality. Flax *(Linum usitatissimum)* has nodes that help the fibers lock together during spinning, but cotton has up to 2,000 natural twists (in alternate directions) to every inch.

Tensile strength, flexibility, and elasticity

Cotton can easily be spun by hand, as here in Ecuador. Similar simple spinning techniques are still employed elsewhere in South America, as well as in Africa and India. The resulting yarn is generally woven into cloth on hand looms.

Mechanized cotton spinning, first introduced in Britain at the beginning of the Industrial Revolution, is now employed throughout the industrialized nations of the world. The factory illustrated *(right)* is in China, which has overtaken traditional producers like India as a manufacturer of cotton textiles and clothing. Neighboring Hong Kong, Taiwan, and Thailand also produce many cotton goods.

Plant products: Vegetable fibers

Sisal, seen here *(left)* being cultivated in Nigeria, is a type of agave *(Agave sisalana)* that is grown for the tough fibers in its long, sword-shaped leaves. Originally from Yucatan, Mexico, the plant is now cultivated in all tropical parts of the world. The fiber, known as sisal hemp, is removed by machine *(right)*, draped over poles, and then left in the sun to dry. It is stronger and more rot-resistant than jute and is used mainly to make ropes, cordage, and twine.

are important if a fiber is to withstand processing stress. Good colorfastness and resistance to decay are needed if it is to be marketable. The newly harvested fiber also has to be quick to clean if it is to be economically produced. Ramie *(Boehmeria nivea)*, for example, has never competed with cotton or linen because it is difficult to clean. The selling price, however, may be largely determined by the amount of moisture the final fabric will absorb. Cotton's tubular structure allows it to take up and release water rapidly, which makes it useful for toweling, comfortable to wear, and receptive to dyes and bleaches.

Cotton—from fiber to fabric

When a cotton boll bursts, four or five tufts of fiber emerge, each containing several seeds; it is then harvested. The fiber, or lint, is sent to a cotton gin to be separated from the seeds and is packed into 480-pound (218-kilogram) bales. When it is ready for processing, the bale is opened, beaten to loosen the fibers, then formed into a sheet, known as a lap, on a scutching machine. Different qualities of cotton are often blended at this stage to ensure a uniform yarn. The roll of fiber is straightened on a high-speed carding machine (a revolving spiked cylinder), which disentangles and cleans the fibers until they form a thin web. The length of time the cotton remains in the carding machine determines the cleanness and quality of the fiber.

Only the longer fibers produce good-quality cotton, so the short fibers (known as noils) are extracted by combing and spun into cheaper yarns. The long fibers emerge as a sliver of strong, smooth material. This sliver passes through a series of rollers, which draw it out into a finer strand and twist it into roving.

Spinning involves elongating and twisting the cotton to the required thickness and length, then transferring it onto bobbins for weaving. The ring frame, invented by John Thorp in the United States in 1828, drafts and twists simultaneously and is used for large quantities. Spinning speeds have recently been significantly increased by the open-ended technique, often referred to as O-E spinning, which uses high-speed rotor disks for twisting the yarn. More delicate qualities are manufactured on a spinning mule, which drafts, spins, and twists in three separate operations. Doubled yarns, for thread and hosiery, are produced by twisting several yarns together. Moistened roving produces a cleaner, more compact thread, but the correct moisture level is crucial.

Three types of yarn are produced: warp, or longitudinal threads; weft, or filling yarn for crosswise lacing, with less twist; and knitting yarns. The warp and weft are woven together on high-speed looms in modern factories. The Sulzer loom allows a wider fabric to be produced and at higher speeds, and it has introduced a more efficient method of weft insertion.

After weaving, further processing is sometimes needed. To increase the luster, strength, and absorbency, some fabric is mercerized.

Cotton fibers take the form of hollow tubes of cellulose, as can readily be seen at the cut ends of the cotton in this electron micrograph of cotton velvet (a fabric with a furry pile made by cutting through small loops formed during the weaving). The fibers are shown enlarged about 400 times.

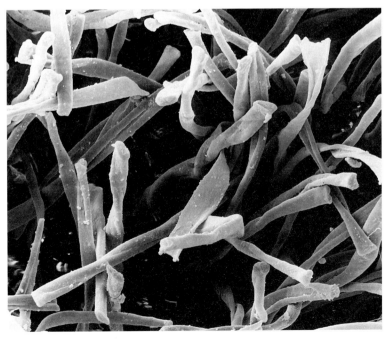

Plant products: Vegetable fibers

In the plants illustrated *(right)*, the leaves or stems are the source of fibers, which are comparatively long. Many are removed from the plants by retting, a process in which the leaves or stems are soaked in water until the fibers can be separated and removed. The softer, short fibers of cotton and kapok are easier to collect because they come from the woolly "hair" on the seeds.

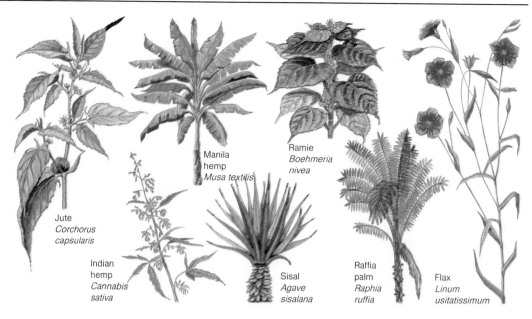

Jute *Corchorus capsularis*

Indian hemp *Cannabis sativa*

Manila hemp *Musa textilis*

Ramie *Boehmeria nivea*

Sisal *Agave sisalana*

Raffia palm *Raphia ruffia*

Flax *Linum usitatissimum*

This technique, devised in 1850 by John Mercer, a self-taught chemist, involves dipping the tensioned cloth into a cold concentrate of caustic soda or liquid ammonia for two to three minutes, then washing it off. Finally, the cloth may be scoured to remove dirt or oil, bleached with sodium hypochlorite or hydrogen peroxide, and colored with synthetic dyes.

The immediate future promises interesting new developments, such as the shuttleless loom, cotton grafted with polymers to produce a cloth with new properties, and greater automation in the finishing stages.

Soft or bast fibers

Bast fibers come from the soft stem or phloem of dicotyledonous plants. They are made up of cell bundles cemented together by nonfibrous gummy substances composed of lignin, pectin, and cellulose. Before the fibers can be extracted from the stem, the gums must be softened, dissolved, then washed away. This process, known as retting, involves immersing the plant stalks in water or exposing them to dew. Bacteria ferments the woody tissues, and enzymes dissolve the binding pectic substances until the fibers can easily be separated.

Known also as linen after its plant of origin, *Linum usitatissimum*, flax is the strongest of the vegetable fibers. Linen garments estimated to be at least 3,500 years old have been recovered from Egyptian tombs. Such resistance to decay and strength made flax the main source of cloth until the arrival of cotton in about 1800. Compared with cotton, however, flax's lack of elasticity caused its popularity to suffer, particularly because it cannot withstand the stresses of high-speed power looms; today it is reserved for high quality, durable, but expensive, household cloth.

China is, by far, the leading country in fiber flax production. Other leading growers include France, Romania, Belarus, and the Netherlands, in that order.

Extracting the fiber

Flax stems are pulled, rather than cut, from the soil by machines that bundle the stems for retting. The better grades of Belgian flax may be retted twice—first for two and one-half days then, after rinsing and drying, for a further day at 90° F. (32° C).

After retting, the stems are split by mechanical breaking rollers and then passed to a turbine scutcher. There the stems are beaten by blades to separate the fibers and crush the pith. The long fibers are combed and all the short fibers removed for coarse-fiber production. Following this so-called hackling process, the long fibers are drawn out through rollers to form a sliver, and then wound on to bobbins for spinning.

Wet spinning is employed for fine linen and warp yarns. The roving is passed through hot water to soften the fibers and to allow them to slip easily over each other. New spinning methods have been devised for flax, including a twistless process that uses the plant's natural pectins to bind the fibers. How-

Jute is grown mainly in the Indian subcontinent. Its uses include the making of carpets *(below)*, which takes advantage of the ease with which it can be dyed a large range of bright colors. Jute is also widely employed for making sacking and burlap.

ever, ring spinning on a flyer machine remains the most economical method.

The flax is then bleached and woven into a range of fabrics, from heavy canvas to fine linen for handkerchiefs. Flax also lends itself to blends with other fibers, such as cotton, Terylene (Dacron), wool, and acrylics, so that the durability of linen can be combined with the lower cost and greater elasticity of other materials.

Hemp

Hemp is extracted from the stem of the well-known drug-producing plant, *Cannabis sativa*. Coarser than flax and very resistant to rotting, it is manufactured into marine cordage, heavy-duty tarpaulins, and, until recently, sailcloth. The higher-quality, water-retted fiber, which turns a creamy white, can be used for finer textiles (one-third of the population of South Korea wears hemp clothing). Unlike flax, hemp is cut in the field after the flowering season. Summer hemp contains better fiber than winter hemp and is normally water-retted to form finer cloth. Hemp is processed in the same way rather than being machine hackled.

Ramie

Also known as China grass and rhea, ramie is the longest, broadest, and one of the strongest textile fibers, but it is very costly to process. Unlike flax, hemp, and jute, it is not retted and the fiber is taken from the stalk by hand. The bark is scraped off in ribbons, which are cleaned in a strong soda solution. They are left to ferment in hydrochloric acid, then returned to the soda to soften and remove the gum. Once the fibers have been extracted, combed, and sorted, ramie is processed in the same way as flax.

Jute

Most jute is white and comes from the Indian herbaceous annual *Corchorus capsularis*. (*Corchorus olitorius* yields the rarer russet-colored tossa jute.) Although jute is the cheapest natural textile fiber and is easy to dye and spin, it is weak, deteriorates quickly as a result of its low cellulose content, and does not bleach well.

The finer qualities are spun into yarn and processed like flax, but 75 per cent of jute is made into burlap, sacking, twine, and carpet backings. As a packaging material, it has difficulty competing with kenaf, roselle, and synthetics such as polypropylene, largely because there has been little mechanical progress in India, the center of the jute industry.

Kenaf

Kenaf, which derives from the Asian plant *Hibiscus cannabis*, is now used increasingly as a substitute for wood pulp. It is stronger, lighter, more water-resistant, and—because the stems are less woody—is better suited to mechanized fiber separation than is jute.

Flowering flax creates a field of blue in the Belgian countryside. Linen, the fabric made from it, was once the only alternative to wool for the people of Europe before cotton was imported from the East and, later, from America.

Alcohol and other chemicals

Ripe grapes *(right)* contain glucose, a simple sugar that can be fermented in the wine-making process to produce ethyl alcohol.

Living plants are biochemical factories. In addition to cellulose (a plant's main structural material used for making paper and rayon), there is a wide range of chemicals produced by various plants, from sugars and fats to complex compounds employed as drugs, dyes, and pigments. Some chemical substances, such as tannins, may be extracted directly from plants. Others, such as alcohols, are made by processing plant materials.

Sugars and starch

The most common sugar in plants is glucose, a product of photosynthesis and the building block, or monomer, from which the natural polymer cellulose is made. Glucose is the principal sugar in grapes. Other simple sugars, known chemically as monosaccharides, that are derived from plants include fructose (in various fruits and in honey) and ribose (in the polymeric backbone of the nucleic acid DNA).

Disaccharides, formed of two monosaccharide units joined together, are more complex sugars. They include maltose (from malt, or sprouted barley) and sucrose (from sugar beet or sugar cane), the sugar used in cooking and for sweetening beverages. This sugar has been used for nearly 8,000 years. It was extracted from sugar cane first in India and much later (after about 300 B.C.) in the Middle East; it was introduced to the Americas by Europeans in 1493. Sugar was not obtained from beets until the eighteenth century. Maple syrup is a sugary sap extracted from various North American species of maple trees (*Acer* spp.).

The chief polysaccharides—polymers made of thousands of monosaccharide (glucose) units—are cellulose and starch. Cellulose and its uses are described in other articles in this section. Starch is a storage material in cereal grains, roots, and tubers such as potatoes. Apart from being the principal ingredient in various kinds of flours, it is also used to make adhesive paste and sizing for textile fabrics.

Vegetable oils and fats

Various plants are grown for the oils contained in their seeds, from palms and olives to flax (the source of linseed oil), sunflower, and specially developed strains of safflower. Vegetable oils are used to make soap and in foodstuffs. One of the more recent additions to the list of oil-bearing plants is soybean (*Glycine* sp.). Since its potential was first realized in the 1930's, annual output has increased more than 200 times, and soy now occupies 68 million acres (27.5 million hectares) of arable land in the United States. The food industry uses most of the oils extracted from soybeans in such

Tannins have long been used for curing animal hides to make leather, either in a traditional way as in Morocco *(below)* or in modern tanneries. A much more recently introduced plant product is soy oil, which is extracted from the seeds of the soybean plant *(below right)* and used in various foodstuffs.

Plant products: Alcohol and other chemicals

Grape-pickers in France work along rows of Pinot vines, the variety employed for making Champagne. Carbon dioxide gas, a by-product of the fermentation process, is retained to give the wine its sparkling, effervescent quality.

products as candy, ice cream, margarine, mayonnaise, and sausages. Other soy oils are used to make glycerin, paints, and soaps.

Numerous other plants yield industrial oils. Citrus seeds, tung (*Aleurites* sp.), and linseed (from flax, *Linum* sp.) provide the bases for enamels, paints, and lacquers; rice bran produces a hard wax and an oil employed in rust-proofing. Linseed and soy oils can be chemically processed to make vinyl ethers, which in turn are polymerized to make tough plastic film.

Safflower *(Carthamus tinctorius)*, originally from India, has been genetically modified so that its seeds yield 40 per cent oil with a linoleic acid content of nearly 80 per cent, hence its popularity as a low-calorie, low-cholesterol ingredient in cooking and salad oils. Recently, new strains have been introduced with a high oleic acid content, making them competitive with olive oil.

Castor *(Ricinus communis)*, originally an ornamental plant from Africa, is now widely grown for its oil-rich "beans." Castor oil is a good lubricant and an ingredient in polyurethane foam plastics, printing inks, and adhesives. Genetic modification has succeeded in creating a uniformly short plant with tripled yields of oil.

Another new and successful product of plant breeding is a member of the Scandinavian mustard family, known as crambe (*Crambe* sp.). Its seeds contain large quantities of erucic acid, which also constitutes up to 50 per cent of rapeseed oil. The substance has found profitable application as a lubricant in the continuous casting method of making steel. It is also used in the processing of rubber and some plastics.

The seeds of a Mexican desert plant called jojoba *(Simmondsia californica)* yield a liquid wax which is used as a "tough" industrial lubricant in place of spermaceti oil from whales. It is also an ingredient of some cosmetics and shampoos.

The flowers of many plants contain oils—extracted by squeezing in presses, using solvents, or by careful steam-distillation—which are the chief aromatic ingredients of perfumes. Also called essential oils, they are the basis of flower-growing industries in several Mediterranean countries. Yields are predictably low—for example, 1.25 tons of rose petals are needed to produce 1 pound of rose oil—and the value of the oils is correspondingly high.

Alcohols

Traditionally, alcohol (specifically ethyl alcohol, or ethanol) is made by fermenting sugars or starch, one of the earliest chemical processes to be discovered. In wine making, yeast is employed to ferment the sugars in fruit; in beer making, the source of the alcohol is the starch in cereals, such as barley. (Any sugary or starchy plant material, however, can be fermented to make alcohol). The fermentation process is accompanied by the evolution of carbon dioxide gas, which is what gives the effervescent quality to beer and some wines.

Most ordinary wines (as opposed to fortified wines such as port and sherry) contain 8 to 15 per cent alcohol, whereas beers seldom have more than 2 to 6 per cent. To extract concentrated alcohol from such sources they have to be distilled (giving 95 per cent ethyl alcohol).

Ethyl alcohol manufactured for commercial purposes is usually "denatured" (to prevent misuse) by adding the highly poisonous substance methyl alcohol (methanol). Also called wood alcohol, it can be made by the destructive distillation of wood—that is, by heating pieces of wood strongly in the absence of air. Methyl alcohol can be added to gasoline to produce "gasohol," a liquid fuel for motor vehicles. Like ethyl alcohol, methyl alcohol is also used as a solvent and as an important starting material in synthetic chemistry, particularly for making monomers for the production of plastics.

Plant products: Alcohol and other chemicals

An unusual liquid wax, incorporated into some cosmetic preparations, is obtained from the crushed "beans" of the Mexican jojoba plant. Jojoba oil is also used as a lubricant for machinery.

Tannins

Since the first animal skins were made into leather, tannins have been used to soften and cure hides. They are complex chemicals called polyphenols, found mainly in the bark and heartwood of hardwoods, such as oaks (*Quercus* spp.) and chestnuts (*Castanea* spp.), as well as in the Australian acacia (*Acacia* sp.) and South American quebracho trees (*Aspidosperma* spp.). Mangrove bark, coniferous trees, the nuts of the betel palm, and the leaves of the tea plant are also rich sources of tannins, whose natural function in the plants prevents decay. They are usually extracted by boiling the vegetable material in water.

In addition to their use in leather making, tannins have many other applications; they are used, for example, in the manufacture of dyes and inks and as coagulants in the early stages of rubber processing.

Drugs from plants

Many drugs occur in the leaves, fruits, seeds, and roots of various plants, and although most can be made synthetically in the laboratory, such is their complexity that it is often more economical to use the natural sources.

Most plant drugs are alkaloids—nitrogen-containing compounds that have an alkaline reaction. Those derived from leaves include atropine, from belladonna *(Atropa belladonna)*; digitalis, from foxgloves (*Digitalis* spp.); cocaine, from the South American coca shrub *(Erythroxylon coca)*; and strychnine, from the evergreen *Strychnos nux-vomica*. People in southeastern Asia have traditionally chewed the leaves of betel-pepper *(Piper betle)*, now the source of a counterirritant drug, and the addictive nature of cigarette smoking has been attributed—at least in part—to the presence of the alkaloid nicotine in the leaves of tobacco *(Nicotiana tabacum)*.

In the case of certain trees it is the bark that is the source of drugs; for instance, the antimalarial drug quinine comes from the cinchona tree *(Cinchona officinalis)*, originally from South America but now also cultivated in Java, and pseudopelletierine, an antihelminthic drug, comes from the pomegranate tree *(Punica granatum)*. In other cases it is the plant's roots that provide the drug; reserpine, for example, which is used for treating high blood pressure, occurs in the roots of the tropical rauwolfia *(Rauwolfia serpentina)*, and ginseng is extracted from the roots of the Chinese panax *(Panax ginseng)*.

The best-known plant whose seeds yield drugs is the poppy *(Papaver somniferum)*, source of opium and its morphine derivatives. Like all alkaloids, the opium compounds are highly poisonous, although the most toxic is probably ricin, extracted from the seeds of a castor plant *(Ricinus sanguineus)*, which has been considered as a possible agent for chemical warfare.

Dyes and pigments

Before the development of the synthetic dyestuffs industry, plants were a major source of colorants for dyes, inks, and paints. Dye plants were grown as farm and plantation crops, and included woad *(Isatis tinctoria)* and indigo (*Indigofera* sp.), whose leaves yield blue dyes, and madder *(Rubia tinctorum)*, which contains

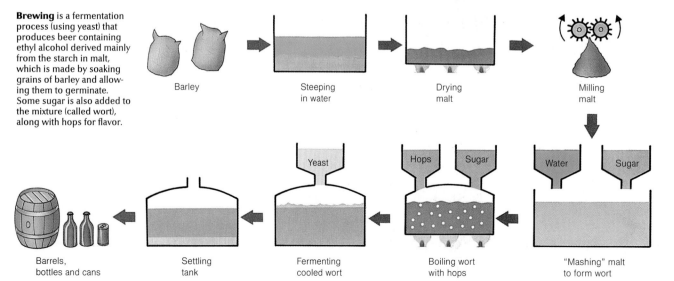

Brewing is a fermentation process (using yeast) that produces beer containing ethyl alcohol derived mainly from the starch in malt, which is made by soaking grains of barley and allowing them to germinate. Some sugar is also added to the mixture (called wort), along with hops for flavor.

Plant products: Alcohol and other chemicals

Sugar cane, resembling bundles of dried sticks, is a difficult crop to handle and requires special machines, although much of the actual cutting is still done by hand. It is now grown in most tropical countries as a source of table sugar and molasses; part of this crop, from Barbados, will be fermented to make rum.

alizarin and various other red dyes in its roots.

The leaves of henna *(Lawsonia inermis)* are still used for dyeing hair and by many Asian peoples for coloring their nails reddish-brown and for decorating the skin. Another red pigment with cosmetic uses (it is employed as rouge) is made from safflower petals. The similarly named saffron, or autumn crocus *(Colchicum autumnale),* yields a yellow pigment employed for dyeing textiles and coloring food.

The chemical indicator litmus—which has a blue color in alkaline solutions and a red color in acid ones—is prepared from various species of lichens, particularly those belonging to the genus *Variolaria*. Lichens also provide soft colors used to dye wool in northern countries.

Chemicals from wood pulp

The liquors that remain after the various processes for preparing wood pulp in papermaking—once discarded as waste—are a rich source of chemicals. In the United States, for example, more than 80 per cent of the annual production of turpentine and rosin is obtained from so-called "black liquor," a by-product of the sulfate process for papermaking. Turpentine is condensed from the vapors and purified by distillation, and the concentrated liquor used as a lubricant and source of rosin, for sizing paper, and as a drying agent in paints and varnishes.

Another product of pulp manufacturing is lignin, which may be burned as a fuel (usually for making steam at the paper mill) or added to clay slurries in the ceramics industry (to reduce their viscosity) or to mud slurries employed as lubricants when drilling oil wells. Lignin can also be hydrogenated—that is, it can be reacted with hydrogen at high temperature and pressure in the presence of a catalyst—to give high-molecular-weight cyclic alcohols used as plasticizers (softeners) for plastics. Vanillin, a flavoring agent that tastes of vanilla, can also be manufactured synthetically from lignin.

The castor oil plant *(left)* has creamy-yellow flowers that give way to prickly capsules containing a large seed or "bean." The seeds are the source of castor oil, used as a lubricant and starting material for making plastics; they also contain the poisonous substance ricin.

An Indian woman's hands display intricate patterns drawn on using the vegetable dye henna, made from the dried leaves of a plant widely grown in India and northern Africa. Sometimes the color is made purple by adding the blue dye indigo.

Rubber

Rubber latex drips into collecting cups from cuts in the bark of rubber trees *(Hevea brasiliensis, far right)*. Almost two-thirds of the world's production of natural rubber comes from trees grown on smallholdings in Malaysia, Indonesia, and Thailand.

Rubber has been one of the most important materials derived from plants since the middle of the nineteenth century. In 1839, the American, Charles Goodyear, invented vulcanization, a process in which raw rubber is blended with sulfur to make it both strong and pliable over a range of temperatures. Only since World War II have synthetic rubbers—products of the petrochemical industry—seriously challenged the natural material.

Today more than 40 per cent of the world's natural rubber comes from Malaysia (most of the rest is grown in Indonesia and Thailand), from smallholdings and plantations growing the South American rubber tree *Hevea brasiliensis*. This plant grows well only within a "rubber belt" that extends about 700 miles (1,100 kilometers) on each side of the equator, and then only at altitudes below 1,000 feet (305 meters). The raw rubber takes the form of a milky latex, which is "tapped" from the trees by making a spiral cut in the bark of the trunk.

The production of natural rubber has increased nearly 95 times since 1900, largely to meet the demand created for motor vehicle tires; the motor industry uses almost 70 per cent of the 4.75 million short tons (4.3 million metric tons) produced each year. But synthetic rubbers take a larger share of the market—60 per cent of the 10 million short tons (9.1 million metric tons) produced annually goes into making tires—although the soaring oil prices of the 1970's helped to reduce the threat to natural rubber.

Composition and properties of rubber

Rubber is a natural polymer, consisting of small molecules of the hydrocarbon isoprene linked together to form long coiled chains. When rubber is stretched and then released, the long-chain molecules bounce back to their original length, giving rubber its elastic properties.

Natural rubber is resistant to alkalis and weak acids but swells in many organic liquids,

such as gasoline and lubricating oils. It dissolves in various volatile organic solvents, producing a rubber solution that finds application as an adhesive. It is a good electrical insulator, although for this purpose rubber has been almost entirely superseded by synthetic plastics. The individual properties of processed rubbers are largely influenced by chemicals added during the manufacturing process.

Processing rubber

Liquid latex, from rubber trees, has to be solidified before it can be sold or further processed. Some solidifies naturally on the tree, but most is coagulated artificially. Any dirt is filtered out and formic (methanoic) acid is added, which causes the liquid latex to "curdle" and coagulate. The coagulum is granulated, dried, and shaped into bales.

About 10 per cent of natural rubber is used in the latex form, where its elasticity and pliability are important for such products as rubber gloves and catheters. The other 90 per cent must first undergo special processes according to what end-product is required.

Before further processing, rubber is masticated by rollers or rotors that break down the polymer molecules. The heat evolved softens the rubber and it becomes plastic. Strength and elasticity are restored by vulcanization

The main steps in the processing of natural rubber into an industrially useful material are chemical: the addition of formic acid to coagulate the liquid latex and, at a later stage, heating with sulfur to bring about vulcanization.

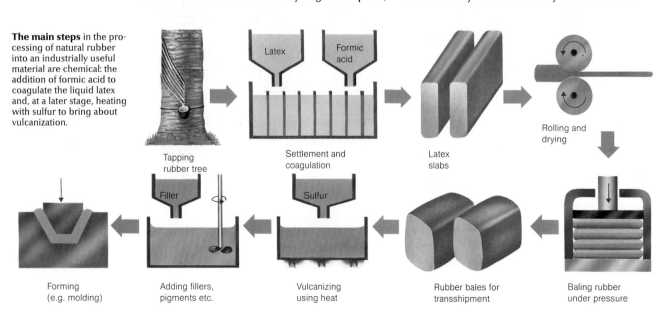

Tapping rubber tree — Settlement and coagulation — Latex slabs — Rolling and drying — Baling rubber under pressure — Rubber bales for transshipment — Vulcanizing using heat — Adding fillers, pigments etc. — Forming (e.g. molding)

with sulfur at 284°-356° F. (140°-180° C). This causes the material to become thermosetting and thus resistant to wide temperature changes. Vulcanized rubber can be stretched up to eight times its original length and yet remain perfectly elastic—that is, it returns to its original length when released.

During vulcanization, the sulfur atoms form cross-linkages between the chains of rubber molecules to produce a more rigid three-dimensional structure. But too much sulfur and, therefore, too many cross-linkages results in a rigid, nonelastic material. Various other substances are also included. Antioxidants are added to prevent the rubber from perishing; oils to soften it and assist processing; insoluble wax to reduce cracking; pigments to supply color; and fillers that, at a slight cost to strength, increase wear resistance. Rubber that is used for car tires, for example, contains 30 per cent of carbon black as a filler. Other products—erasers and floor tiles, for instance—may have white powder fillers such as chalk, silica or zinc oxide.

Shaping rubber

The usual method of making dry rubber goods is by molding. A simple eraser, for example, is heated under pressure in a mold so that it emerges both vulcanized and shaped. For more precision and higher production rates, injection molding is used. In this process a softened rubber mixture is injected at high speed through small holes into a mold or set of molds. Heat and pressure are again applied to vulcanize it. Surgical rubber goods, such as thin latex gloves, are produced by dipping shaped formers into precompounded liquid latex, and then drying and vulcanizing the products.

Rubber hoses, tubes, and hollow casings are formed by extrusion. The rubber mixture is forced through a die, rather like squeezing toothpaste from a tube, and vulcanized using hot air. Rubber sheeting (used for weather balloons), fabric-reinforced sheeting (used, for example, on hovercraft skirts), and proofed fabric (used for raincoats) are all formed on a calendar. This is a machine with rollers that squeeze the rubber into a sheet. To produce a waterproof material, fabric is fed through the lower rollers so that the rubber sheet is pressed into it.

An inflatable rubber boat carrying 11 men *(above)* shoots the rapids on a fast-flowing river, testifying to the strength and resilience of rubber. The scientist *(below)* uses polarized light to study the stress patterns in a motor vehicle tire. More than two-thirds of the natural rubber produced goes into making tires for the automotive industry.

The French cotton plant *(Calotropis procera)* is a flowering herbaceous plant that produces latex. When a leaf is snapped, the latex flows freely from the plant tissue.

Latex and other plant gums

Latex is a milky-white liquid that occurs in the cells of some flowering plants. It is a complex mixture of substances that may include fats, waxes, and various gum resins. It circulates the plant tissue like sap in branched tubes, conducting other substances and acting as an excretory reservoir. Latex is produced especially by plants of the family Asclepiadaceae, but also by those in the Sapotaceae, Euphorbiaceae, and several others.

Gums, unlike latexes, are clear, yellow, or amber liquids that harden into translucent solids when exposed to the air. The families Leguminosae, Rosaceae, and Sterculiaceae contain most of the gum-producing trees, which convert cell tissue into gum, probably by enzymic action. In some species, the gum collects in ducts to heal damaged tissue and, in others, such as gum arabic *(Acacia senegal)*, oozes from the bark in response to injury.

Resins are obtained from the sap of various plants; like latex and gums, they consist mainly of hydrocarbons based on isoprene, and their main chemical compound is abietic acid. They may be tapped straight from the tree or collected from the ground, sometimes as fossilized lumps (amber). Shellac is produced from the resinous excretion of a scale insect *(Laccifer lacca)* that feeds on various resin-containing trees, such as the soapberry *(Sapindus* sp.) and acacia *(Acacia* sp.) trees of India and Burma. Most resin, however, is tapped from pines, particularly the longleaf pine *(Pinus palustris)*, which grows in the Southeastern United States.

Natural latex

Most of the commercially used latex is extracted from the Pará rubber tree *(Hevea brasiliensis)* and is processed into rubber, the subject of the previous article. Other important products derived from natural latex include gutta-percha, balata, and chicle.

Gutta-percha is a yellow or brown leathery material from the latex that is produced primarily by the *Palaquium oblongifolia* tree. This tree grows wild and is cultivated on plantations in Malaysia, the South Pacific, and South America. To extract the latex, trees may be felled or rings cut in the bark. On plantations, fresh leaves are chopped, crushed, and boiled in water to remove the latex.

Gutta-percha becomes plastic and water resistant when it is heated. It can, thus, be used as an insulating material. It is employed for underwater electrical equipment as well as for golfballs and chewing gum. Development in synthetic materials, however, has meant that the use of gutta-percha has largely been replaced by products such as polyethylene, nylon, and vinyl resins.

Balata is a hard, rubberlike material made from the milky juice produced by some tropical plants, notably the bully tree *(Manilkara bidentata)*. The trees are grown particularly in

The llareta *(Azorella glabra)* is a ground-hugging cushion plant that grows high up in the Bolivian Andes. It exudes drops of resin that harden when exposed to the air. Bolivians use the resin as fuel because it produces intense heat when burned.

Guyana and the West Indies. Balata is often used as a substitute for gutta-percha to make golfballs and belting.

Chicle is said to contain both rubber and gutta-percha and is a pink or brown material consisting of partly evaporated milky sap. The sap is obtained mainly from the *Achras zapota* tree, which grows in the West Indies, Mexico, and Central America. Chicle was originally developed as a base material for chewing gum, but has been largely replaced by synthetic products.

Some herbaceous plants produce latex, such as the rubber dandelion *(Taraxacum koksaghyz)*. The roots of this plant are composed of 8 to 10 per cent rubber.

Plant gums

Gums are soluble in organic solvents, such as alcohol and ether, but they form a gelatinous paste when steeped in water—qualities that have found applications in many industries. The most widely used gum is arabica, which is extracted from the *Acacia senegal*. It consists mainly of calcium, potassium, and magnesium salts of arabin (a complex polysaccharide, which, on hydrolysis, yields glucovonic acid and various simple sugars).

Gum arabic has many applications, the most familiar of which are as an adhesive and as the main glue used for postage stamps. It is employed in surgery to bind severed nerves and will maintain blood pressure because it matches the osmotic pressure of blood. The printing industry is a major beneficiary of gum arabic, particularly in lithographic processes where it acts as a demulcent in the preparation of chemical emulsions. This property also makes it useful for thickening inks.

Large quantities of gum are consumed in food and are used in pharmaceutical products. Its sticky texture makes it an excellent binding and thickening agent for lozenges, pills, and candy, and its pleasant smell and consistency also render it a popular additive in a variety of foods and cosmetics. Gum benzoin, which is obtained mainly from the snowbell tree *(Styrax benzoin)*, is used in these products, as is gum tragacanth, which is the dried exudation of the milk vetch (*Astralagus* sp.).

Resins are used to make a diverse range of products, from paints and varnishes to turpentine and perfume. The pine essence of some resins adds fragrance to numerous household products and the distinctive taste to Greek retsina wine. Turpentine, or oil of turpentine, is an aromatic liquid made by distilling resin. It is employed as a solvent and as a vehicle or thinner for oil paints. Some resins have antiseptic properties and are added to cough syrups and mouthwashes.

Amber—a yellowish, usually translucent, fossilized resin from the extinct pine tree *Pinites succinifera,* which once grew along the Baltic coast of Europe—softens when heated and can be fashioned into ornaments and jewelry. It also generates static electricity when rubbed, a property that (from the Greek *elektron,* meaning amber) gave rise to the word electricity. Amber is also produced by tropical trees in Central and South America.

Rosin is the residue from the distillation of the volatile oils of resin. As well as being used for paints and varnishes, it is also applied to violin bows and, in a powdered form, is used by ballet dancers and gymnasts to prevent them from slipping.

Camphor is a resinous material that is extracted from the camphor tree *(Cinnamomum camphora),* which grows in Indonesia, China, and Japan. It is steam-distilled from the chopped-up wood of trees, which must be at least 50 years old. Like other resinous products, camphor is used in lacquers and varnishes, but it is also an ingredient in the manufacture of some moth-repellents and explosives.

In Portugal, the maritime pine *(Pinus pinaster)* is tapped for resin. Oil of turpentine is made from distilled resin and is used as a thinner for oil paints and as a solvent.

Amber is fossilized resin and occurs naturally as irregular nodules. It is usually orange or brown in color and may be opaque. This piece of amber *(above)* was found in the Baltic region of Poland. The insect inside was trapped before the resin hardened.

Conservation and reclamation

The number of vascular plant species considered to be in danger of extinction is roughly estimated to be about one-tenth of the total number of plant species in the world. It is not just the extinctions that are a cause of concern, but the time scale within which they are occurring. Plants cannot evolve fast enough to withstand habitat destruction by, for example, bulldozing, forest cutting, overgrazing, or damage by pollutants. Indeed, the problem is growing ever more critical as the human population on earth increases.

The plant world is essential to all animal life, including humans, as the base of the food chain. All our food, from cereal crops to grazing livestock, is directly or indirectly a plant product. In addition, plants protect soil from erosion and, frequently, nourish it; they also play a major role in the world's climate and carbon dioxide balance.

The three major arguments for conservation are economic, climatic, and esthetic. Of them all, the economic one has the greatest immediate impact. This argument is also the strongest in justifying land reclamation. If species are to be conserved in their original habitats, then less land must be developed to cope with human population expansion and the consequent agricultural needs.

Economics of conservation

In the United States it has been estimated that nearly one-quarter of all drugs obtained on prescription in any year are obtained from plants. They include tincture of arnica (from mountain arnica, *Arnica montana*), atropine (from belladonna, *Atropa belladonna*), codeine and morphine (from the opium poppy, *Papaver somniferum*), and quinine (from the bark of the *Cinchona* tree). About 80 per cent of major drugs probably can be derived from plants more cheaply than they can be made in the laboratory.

But many plant species that are potential sources of drugs are in danger of becoming extinct before we learn to exploit their special properties. One example is a member of the periwinkle family Apocynaceae, *Catharanthus coriaceus,* which is found in Madagascar in very small numbers and is threatened by grazing and burning. The *Catharanthus coriaceus* plant could be of great use although its medical value has not yet been thoroughly investigated—its genus is known to contain 70 alkaloids, some with clinical value in the treatment of cancer.

Plants provide many products other than drugs, including rubber and gums (*see* previous articles), tanning agents, dyes, fibers, insecticides, perfumes, waxes, cosmetics, preservatives, turpentine, and oils. This last product has seen an interesting development: sperm whale oil was used from the 1930's for several years as a high-pressure lubricant and had no equal. By the late 1960's these whales

Cattle now graze on Montana land reclaimed after strip coal mining.

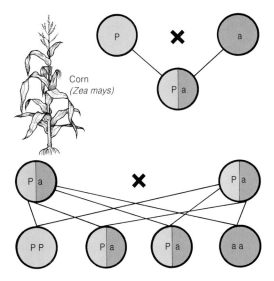

were greatly reduced in number and approaching extinction. Fortunately, it had been discovered that oil from the jojoba bean *(Simmondsia californica)* can replace the sperm whale oil as a lubricant and so reduce the need to hunt these animals, which are now protected.

Forests, being the richest sources of timber, are being exploited everywhere. For example, the equivalent of several square miles of forest is cut down in Malaysia every day for export. In 1950, 35 per cent of the earth's land surface was covered in forest; today it is about 30 per cent. Apart from supplying the timber industry, wood is still a major source of fuel—it provides 31 per cent of the world's fuel needs and plays a large role as a source of energy, mainly in developing countries. But in many forests, once the plant cover is removed, the soil is easily washed away, and the ground is left infertile.

Unlike fossil fuels, wood is renewable; it can also be converted to various other fuels and, thus, be competitive in a world economy. Controlled and artificial regeneration of trees is, therefore, desperately needed to ensure that stocks of timber will be replenished for future generations and that, with care, forests will always provide what is required.

Yet another economic need is that of soil conservation. The removal of plant cover almost inevitably results in soil erosion to some degree; for example, heavy rains can gradually turn a deforested hillside into a bare rocky slope, washing away all the soil and depositing it in rivers or lakes. In river watersheds this erosion can cause rivers to silt up; similarly, reservoirs and lakes may silt up, preventing their use for either water supply or hydroelectricity. But it is not only in watersheds that the removal of the native vegetation brings problems of erosion. The removal of plant cover on plains may lead to wind erosion of the soil and the creation of dust bowls, similar to what happened in the United States during the 1930's, when farmland of the Middle West was damaged. Conversely, however, it is often possible to reduce erosion by introducing plants to protect the soil from the effects of wind and rain.

Pets and domestic animals have also caused a great deal of destruction. Grazing may prevent seed formation, and overgrazing can kill off plants or at least keep them at subsistence level so that they die without seeding or even managing to reproduce vegetatively. The presence of grazing animals, such as goats, on small islands with endemic plant species has had a drastic effect on the plant life there. Proper grazing rotation could conserve some grasslands as pasture and is now practiced in certain areas. Controlled burning is encouraged in these areas because it causes the leaf litter to release its nutrients and thus help to increase nutrient levels in the soil, as well as to stimulate new plant growth. The addition of artificial fertilizers also helps to maintain the vegetation.

Conservation for food

Food must also take its place in any economic consideration of conservation. About 3,000 plant species have been recorded as providing humans with food, but most present-day agriculture is based on fewer than 30 species. Of these, three plants are of particular importance: rice *(Oryza sativa)*, wheat *(Triticum* sp.), and corn *(Zea mays)*; about half of all arable land in the world is taken up with their cultivation. The dependence on such a small number of crops is potentially disastrous, because monocultures are more susceptible to disease and climate variation than mixed crops.

Hybridization often uses the gene pool in wild crop plants to improve cultivated varieties. Research has been done into the possibility of breeding a perennial variety of corn *(Zea mays, above left)*. The dominant gene (P) in the wild perennial variety is bred with the recessive gene (a) in the cultivated annual variety. When the hybrids (Pa) cross-fertilize, three out of four will be perennial. Similarly, rice *(Oryza* sp.), which occurs naturally in tropical and subtropical regions, can now be grown in Portugal *(above)*, where it provides a staple crop, having been bred to withstand temperate conditions. The need to conserve wild plants for their genetic importance is slowly being acknowledged.

136 Conservation and reclamation

Land reclaimed from the sea has for centuries been turned over to pasture, arable farming, or horticulture. One of the largest such reclamation projects—the Zuider Zee scheme—has been undertaken in the Netherlands. Over the decades, this country, which covers 16,163 square miles (41,863 square kilometers), has reclaimed about 3 per cent of that area.

Some potential crop plants, however, are threatened species, such as the Yeheb bush *(Cordeauxia edulis),* which grows on the Somali-Ethiopian border and is being killed off by overgrazing. It produces nuts that have a very high protein content and are, therefore, a valuable food source. It can grow in the most arid parts where other legumes cannot survive, and, once planted, it needs no tending. It is, therefore, very suitable as a crop for the nomadic populations of the area.

These and other wild crop plants bear the characteristics that make it possible to develop new and better varieties of crops by breeding in disease-resistance or greater yield. This is done genetically using a repository of genes (a gene pool). These gene pools survive in places such as uncultivated grasslands and rain forests that are of inestimable value as a gene bank—a value which disappears as these areas are destroyed.

Climatic conservation

The presence of vegetation and its removal is thought to affect the climate. A forest, for example, is continually giving off water vapor into the air that may affect the amount of rainfall over the forest. Its destruction may have more than just a local effect on rainfall. There may be, for this reason, a desperate need to conserve the world's rain forests, which are being irrevocably destroyed.

Plants also remove carbon dioxide from the atmosphere and store it. After deforestation, and the loss of humus by erosion, carbon dioxide is released into the atmosphere and accumulates there, leading to a "greenhouse effect." It is thought that if the carbon dioxide content of the atmosphere is raised by even a small amount it will cause the average annual temperature to rise by a few degrees. This increase in temperature could melt the polar icecaps, which would raise the sea levels and cause climatic conditions to alter.

At present, 22,000 million tons of carbon dioxide are added to the atmosphere every year. Afforestation and the conservation of forests is needed to remove and store it. It has, therefore, been suggested that large-scale afforestation of fast-growing trees be undertaken and, indeed, there are over 250 million acres (101 million hectares) of newly created forest today.

Another alarming fact that impresses on us the need for plant conservation is that 50 to 60 per cent of the oxygen on this planet is produced by terrestrial and freshwater plants (the balance comes from marine algae). Were this source to be depleted, it would have serious consequences for living organisms.

Esthetic conservation

There is no doubt that plants are visually appealing to many people and, while satisfying horticultural needs, can often be combined with recreational areas. Forests and gardens, for example, have been places of leisure activity for centuries. The commercial value of forests need not be reduced by conservation or recreational needs. Botanic gardens and nature reserves also serve as reservoirs for endangered species, in which they can maintain genetic variability and act as a gene pool for future hybrids. Wildlife is closely associated with vegetation and in conserving large tracts of vegetation, we maintain wildlife habitat.

Reclamation

The conservation of plants is the first step in preserving our plant world, but there is also a pressing need for more space for expanding human populations, their food, and that of their livestock. Some space can be retrieved from coastal dunes and marshes, deserts, and disused quarries and mine dumps.

Land has been reclaimed from the sea for centuries. These days, the Dutch method of empoldering is most often employed in western Europe. Sediments formed from alluvial gravel, sand, silt, and decaying vegetation are carried in by the tide, accumulate behind artificial dikes, and the enclosed land is drained. Gradually rain leaches most of the salt out of

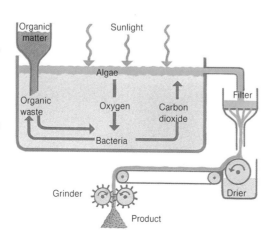

Algae farming is an alternative means of producing feed and fertilizers and indirectly may protect plants from exploitation. Bacteria in ponds feed on waste, producing carbon dioxide that is used by algae in the water, together with sunlight, in photosynthesis. The algae are skimmed off the water, filtered, dried, and then used.

the highly fertile soil, which can then be used for agriculture. In Germany in the 20th century alone, about 30,000 acres (12,000 hectares) of marine alluvium have been reclaimed in this way.

In tropical and equatorial regions, however, many marshes are high in sulfides, which when exposed to the air become sulfates, making the soil too acid for crop production. In southeastern Asia and India, rice is grown economically on such soils by not totally draining the land and avoiding the use of heavy machinery on it, which would break up the soil.

Coastal and desert dunes can be stabilized in several ways—for example, by planting grasses or succulents on them, by spraying resins onto the sand surface, or by covering it with polyethylene sheeting. Once stabilized they can be planted with crops, with trees for timber, or even forage plants. In the Wellington province of New Zealand, for example, pampas grass *(Cortaderia selloanna)* has been grown on stabilized sand as a food reserve for cattle. Desert reclamation has combined stabilization of the sand with various methods of irrigation and the addition of fertilizers and chemicals.

Apart from shores and deserts, some areas of land devastated by industrial machinery and strip mining have also been reclaimed. In parts of the Australian desert, for example, the sand has been heavily mined for mineral salts, and the soil structure has consequently been destroyed. The toxic mining wastes reduced the possibility of natural reclamation of the area, but with the application of at least 4 inches (10 centimeters) of topsoil vegetation could be established. Banksia seedlings were planted together with sorghum to hold the topsoil. Nitrogen and phosphate fertilizers were applied, and wattles *(Acacia* spp.) and gums *(Eucalyptus* spp.) were then planted.

The reclamation of land for recreational purposes is an increasing trend. Here, for example, disused gravel pits have been filled with water to provide facilities for sailing and other water sports. This is a suitable reuse of land that has been disturbed by quarrying and that is not easily converted to agricultural use. In a small way it meets the demands of expanding populations.

Forest management

Given good management, trees are a renewable natural resource that provide many products including fuel, lumber, wood pulp, fiber, and food for human and animal consumption in the form of fruits, bark, and leaves. Forests are also ecologically valuable because they recycle nutrients, fix energy, retain soil and water, and keep the soil fertile. They may also influence rainfall and climate generally.

The problem is that pressure from the need for space for cash crops and felling for lumber has led to large-scale destruction, particularly of tropical rain forests, which are a major source of hardwood timber for the world. Scientists estimate that tropical deforestation wipes out about 7,500 species a year.

Forests are delicate systems which, if disturbed, can be permanently destroyed, as has been the case with many tropical forests. It is, therefore, in our own interests to extend, conserve, and manage the forests of the world rather than simply exploit them for short-term gain. Proper scientific management of forests (silviculture) includes the establishment, development, and reproduction of trees to provide salable lumber in the shortest possible time, to control erosion, protect watersheds, provide recreation and enhance the landscape, to protect animals, and to make provision for agriculture.

Forest establishment

Afforestation is the primary establishment of forests in previously unforested areas or those that have long been deforested. To begin, the ground needs to be broken up and drained and, if sloping, may also be terraced to prevent runoff of water. Soil and fertilizers, such as phosphate, are then applied in the planting holes. The soil, climate, quality, and type of timber expected are considered, but the trees chosen are not always indigenous; more importantly they need to be fast-growing and hardy: for example, Sitka spruce (*Picea sitchenis*), which is used in Britain; Monterey pine (*Pinus radiata*), which is widely grown in warm temperate climates, most extensively in New Zealand; and Caribbean pine (*Pinus caribaea*), which is planted throughout the seasonal tropics and subtropics. The fast-growing pines are used for pulp. Other fast-growing trees make veneer logs, such as *Gmelina arborea,* which is grown in Brazil. With good management, trees take up to 50 years to mature; to attain saw log size, a mature forest ecosystem may take twice as long to reestablish.

Seedlings are first raised in a nursery where they may have been bred for hardiness, fast growth, or dense or soft wood. They are usually planted out when one to four years old, during their dormant period. They can be planted at other times provided they are guaranteed water to prevent them from drying out before the roots establish themselves—during a tropical rainy season, for example. After planting, the trees are weeded for the first few years until they are tall enough to overtop and suppress any weeds.

The number of trees planted in a given area depends on the species and the planting site. In temperate areas, hardwoods such as ash (*Fraxinus* sp.), beech (*Fagus* sp.), and oak (*Quercus* sp.) are usually planted 5 feet (1.5 meters) apart. Conifers such as larch (*Larix* sp.) and pines (*Pinus* spp.) are usually planted 5 to 6.5 feet (1.5 to 2 meters) apart, whereas poplars (*Populus* spp.) are usually planted 16.5 to 19 feet (5 to 5.8 meters) apart.

When the developing trees are about 8 to 10 years old the forest floor is generally cleared of bushes and woody climbers. The low branches of the trees are removed, which helps to reduce the number of knots in the mature timber. The trees are thinned out about 12 to 15 years after planting, the thin-

Spruce saplings, in the foreground, are grown for 10 years in the European mountains, then are cut for pulpwood. Spruce wood is thought to make some of the best paper pulp. These trees grow faster and on poorer soils than do broadleaved species.

Limbing, the removal of the lower branches of trees when they are about 10 years old, serves to reduce the number of knots in the mature timber. The removed young wood is then used for pulp.

nings providing the first lumber crop. Their removal allows the final crop to achieve its maximum potential. Diseased and unsatisfactory trees are also removed at the same time.

In caring for forests, such factors as frost, snow, lightning, sunlight, water, and wind need to be considered. Measures taken to reduce the damage done by these elements include proper drainage, adequate provision of water, and shading seedlings. Fires are often the major hazard, causing damage that varies from slight bark scorching to complete destruction. Surface fires, if not too intense, may be beneficial. Such fires are deliberately started in mature eucalyptus (*Eucalyptus* sp.) forests in Australia and in pine forests in the Southeastern United States, for example, where they prevent the build-up of a deep flammable litter layer so that the intensity of fires is kept low. Many species of gum need fires to trigger germination. To prevent fires from spreading, fire lines 33 feet (10 meters) or more across are created and kept clear of vegetation.

Monocultures versus mixed forests

Natural forests usually contain a mixture of tree species. However, many forestry practices have favored single species, for these forests take less time to plant, require less skill to thin, yield more uniform timber, and are more economical to harvest. Ecologically mixed forests are more desirable than monocultures because different species have different requirements for growing conditions and are less susceptible to drought and disease.

Mixed forests are created either by natural succession or by planting several species during reforestation. In Europe, ash *(Fraxinus)* and beach *(Fagus)* may be planted with one or more coniferous species. In the Northwestern United States, red alder *(Alnus rubra)* often establishes along with Douglas fir *(Pseudotsuga menziesii)* and western hemlock *(Tsuga heterophylla)*. The alder and hemlock do not live as long as the Douglas fir, nor is their wood as valuable.

Pests and diseases are a problem, especially with monocultures, among which they spread quicker than in mixed forests. Aphids and sawflies, for example, create galls, and mites, moth larvae, scale insects, and some fungi can cause defoliation. Other fungi damage the wood, fruits, and roots, and eventually kill the trees. The application of insecticides, such as nicotine and malathion, fungicides, tar-oil washes, copper compounds, and burning are effective in reducing disease. However, these measures can at the same time damage the trees and kill off other animal life. Natural control of these pests is, therefore, preferred—insect-eating birds and other animals can provide some measure of control. Animal life is, however, more abundant in natural mixed forests than in monospecific plantations.

Lumber extraction

In the planted forest everything is organized to extract lumber as economically as possible, which is usually by removing all the trees and then replanting the whole area. This method is known as clear-felling or clear-cutting. The area to be felled is calculated by dividing the total area of forest by its rotation time (the time a species needs to grow before it can be cut). If, for instance, a total forest is 500 acres (200 hectares) with a rotation of 100 years, then the area cut and replanted each year is 5 acres (2 hectares). The rotation times vary depending on the species; oak requires 100 to 150 years, and poplar, 40 to 45 years. For a fast-growing tropical tree grown for pulp, less than 40 years may suffice.

The advantages of clear-felling are that it is the simplest system to manage, no seedling trees are damaged during felling, and replanting is more dependable than natural regeneration (even if it is more expensive). If, however, planting is not started soon after felling, weedy species may establish. In addition, the absence of mature trees to provide cover for young seedlings may slow down their establishment.

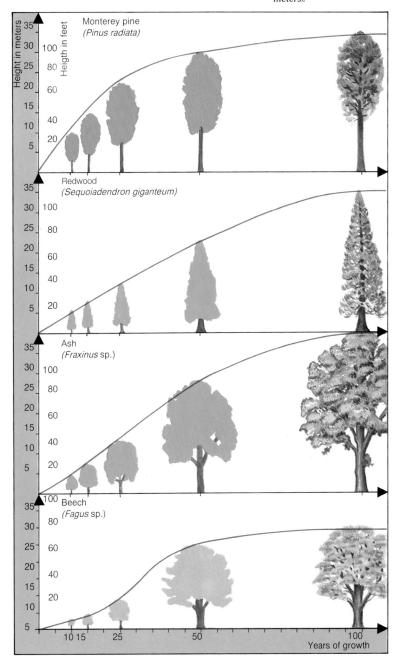

The growth rates of forest trees vary depending on the species. Of the softwoods, Monterey pine can be one of the fastest growers, sometimes shooting up as much as 18 feet (5.5 meters) in one year. Redwood grows more steadily, slowing down as it reaches 100 feet (30 meters). Among the hardwoods, ash grows fast, often reaching 100 feet (30 meters) well before it is 100 years old. It achieves a mature height of about 150 feet (45 meters). In contrast, beech grows very slowly to begin with but later spurts up to reach about 85 feet (26 meters).

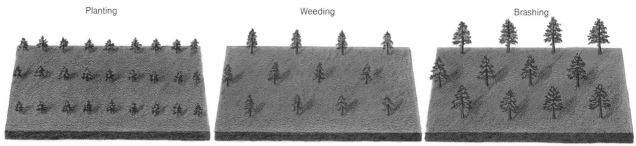

The management of a pine forest grown for timber starts when the saplings, of between two and four years old, are planted. From four to seven years old they are weeded and then, at about eight to ten years, the lower branches are cut off for pulp. Two years later they are thinned, the wood also being used for pulp. They are felled all at once when mature, and soon after the ground is replanted.

Fires are often deliberately caused in Australian eucalyptus forests to limit the undergrowth and initiate seed germination. If the understory were allowed to become too dense and were set alight, the fires would be less controllable, becoming crown rather than ground fires.

A mature natural forest presents a slightly different problem in lumber extraction. A selective system may be operated, in which trees are felled singly, in twos and threes, or in small areas of up to one-third of an acre (in a temperate forest). The trees are felled when mature or when commercially valuable.

The advantages of this system over clear-felling are numerous: natural regeneration is the norm after selective felling and is supplemented by planting only where necessary; the soil is not exposed, so there are no problems of erosion or landslip; and seedlings are protected by the surrounding trees. Moreover, the financial returns are immediate because the trees are already mature. The disadvantages of selective felling are the difficulties and extra cost of extraction and the damage that may be done to the young regenerating trees.

Forests can be felled in strips instead of a block, which simplifies the problems of extraction. Cut at right angles to the prevailing wind in the lee of the wood, the strip is sheltered and can be seeded from the standing trees. When regeneration is underway the next strip can be cut. This is repeated until the whole mature forest is cleared.

Another variant of selective felling is the polycyclic system in which selected mature trees are felled to open up the canopy, which allows previously shaded seedlings to grow and develop. A second felling thins out the canopy further to allow the regeneration process to continue. The final felling removes all the mature trees. The advantages and disadvantages are similar to those of single group and strip-felling.

Coppicing techniques

Coppicing systems have been used in Europe for hundreds of years and provide a mixture of large and small hardwood timber. Because it relies on regrowth from a cut stump, however, this method cannot be used for conifers, which do not normally regenerate from a stump.

A coppice wood is composed of a number of stumps in various stages of regeneration that are managed on a short rotation, dependent on the species. For ash the rotation time can be as little as 5 years, for alder (*Alnus* sp.) it can be 25 years. Initially a mature tree is felled to leave a stump, or stool, 1.5 feet (46 centimeters) high. This is allowed to regener-

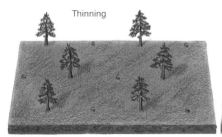
Thinning

Felling

Replanting

ate, and the new shoots grow until they reach the correct cutting size. The shoots grow quickly, nourished by the mature root system. The coppice is generally cut in strips or blocks, the area to be cut being calculated by dividing the size of the coppice by its rotation time.

A number of species are suitable for this treatment and include hazel (*Corylus* sp.), sweet chestnut (*Castanea* sp.), hornbeam (*Carpinus* sp.), and birch (*Betula* sp.). The life of the stools varies depending on the species—for oak it is more than 100 years.

If the coppice is combined with tall trees, the system is known as coppice-with-standard. The stools are mixed with "standard" trees which are typically oak. The standards are normally felled on a rotation that is, for convenience, a multiple of the stools' rotation; they may themselves be regenerated by selecting and retaining one shoot from a coppice stool, rather than grown as "maidens" from seedlings.

Other factors of management

Considerations of erosion and watershed management are important in forestry. Large-scale clear-cutting can cause erosion, waterway and reservoir silting, damage to fisheries, and loss of drinkable or usable water. Additionally, with clear-cutting there is an increase in floods in regions with high precipitation, such as the Pacific Northwest. Good forest management by the regular opening up of the forest canopy increases the diversity of forest structure and consequently that of the flora and fauna.

Population pressures and increasing demands for space for crop cultivation have led in some places to a need to devise management programs that allow the production of food and wood from the same land. Mixed cropping and grazing have also been implemented in forestry plantations.

The potential of forestry as a source of rural development, rehabilitation of degraded land, water catchment, and provision of shelter and recreation is now being appreciated. So too is the need to safeguard the diversity of the species-rich natural forest ecosystems—not simply for the sake of diversity, but for the sake of life on earth.

Abundant, vertical slim branches are the distinctive mark of coppiced trees. The low thick stump, or stool, in the foreground regenerates new shoots every time the old ones are cut for timber.

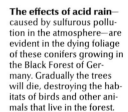

The effects of acid rain—caused by sulfurous pollution in the atmosphere—are evident in the dying foliage of these conifers growing in the Black Forest of Germany. Gradually the trees will die, destroying the habitats of birds and other animals that live in the forest.

Shore and desert reclamation

Most coastal plains are rich in plant nutrients and are, therefore, suitable for agriculture. But a combination of human activities, overgrazing, and erosion can reduce this potential. With enough water for irrigation, shores and even deserts can also be used for cultivation, although the loose composition of sandy soil (and its large grain size) causes rain water to soak through it rapidly, leaching out the nutrients. Such terrain is also particularly vulnerable to wind erosion, which tends to strip away the topsoil. For thousands of years people have tried to reclaim coastal and desert land, to stabilize the areas, and eventually cultivate them.

Shores and dunes

Estuaries and shorelines are reclaimed initially to protect the land from tidal flooding and the seepage of salt water.

Coastal dune systems are also formed to protect the hinterland. They usually comprise a frontal dune, which forms a buffer zone, and a hind dune, which protects the sheltered areas from salt spray or sand-drift damage. Dune formation can be encouraged by placing an obstacle in the path of the prevailing wind, which causes it to slow down and deposit some of the sand that it is carrying. The obstacles may be fences but most often consist of plants. Marram grass *(Ammophila breviligulata)*, for example, has a rapid rate of upward growth, is perennial, and can reproduce vegetatively. It therefore increases the rate of sand accumulation while keeping above the rising levels. Dunes produced in this way typically reach a width of 100 feet (30 meters) or more and a height of 33 feet (10 meters) in about 10 years. Other dune-forming methods include the spreading of a film of rubber compound on the mobile sand, which stabilizes it enough to allow the germination and growth of such grasses as couch grass *(Agropyron junceiforme)*.

Once the initial stabilization of the dune has been achieved, tree lupines *(Lupinus arboreus)* are frequently planted to further stability, together with the spiny sea buckthorn *(Hippophae rhamnoides)*, which discourages animals and people from trampling on the dune surface, causing it to erode.

Dunes are made usable in several ways after stabilization. Nitrogen fertilizer is usually added, and shrubs such as Scotch broom *(Cystisus scoparius)* are planted as an intermediate step before trees are established. Crops of lucerne or alfalfa *(Medicago sativa)* are planted in temperate areas to introduce nitrogen into the soil. Two to four years after planting marram grass, the dunes are sufficiently stabilized to introduce tree species that can tolerate salty winds, such as Monterey pines *(Pinus radiata)*, gum trees *(Eucalyptus* spp.*)*, and acacias *(Acacia* spp.*)*. These plants are deep-rooted and can resist drought. They also grow rapidly and can soon form a stand of mature trees. Acacias help to increase the nitrogen content of the soil by means of bacteria contained in their root-nodules.

In Argentina, crops of rye and sorghum, with the addition of millet, have successfully stabilized dunes covering 15 acres (6 hectares) in as little as 18 months. This has then facilitated the spread of natural vegetation.

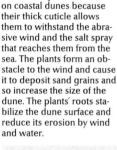

Succulent plants survive on coastal dunes because their thick cuticle allows them to withstand the abrasive wind and the salt spray that reaches them from the sea. The plants form an obstacle to the wind and cause it to deposit sand grains and so increase the size of the dune. The plants' roots stabilize the dune surface and reduce its erosion by wind and water.

Deserts

The main cause of desertification is inappropriate methods of land use on the fragile semi-arid marginal lands.

Desertification can be reduced by the rational use of land resources from the individual level upward, by the introduction and improvement of proper irrigation methods, increased soil moisture storage, the restoration of degraded pastures, and the improvement of strategies of pasture rotation.

Some deserts have underground water deposits that have been tapped by wells to enable the irrigation of crops. This practice is used in north Africa, the Middle East, and the southwestern deserts in the United States. In the American deserts, water is being used more rapidly than it can be replaced. The result is the need to drill deeper wells for the limited water.

One of the problems of irrigation is that, depending on the mineral composition of the water, it can lead to the salinization or alkalinization of the soil, and vegetation cannot be established. To counteract salinization, flushing irrigation is used to keep the soil constantly moist; the careful application of chemicals to the soil, such as gypsum, prevents alkalinization.

Pastures that have been overgrazed have been improved in some desert areas, as in Kazakhstan, by the introduction of forage plants. These include prostrate summer cypress *(Kochia prostrata)*, oriental saltwort *(Salsola orientalis)*, and Mexican mesquite *(Prosopis* sp.). The value of these shrubs is that when grown in stands they reduce wind speed and protect the soil from drying out and contracting. In addition, they survive the parching summer heat (whereas many other plants die down) and so provide a food source in summer. They also have a richer protein content than many grasses do.

The use of sand covers, such as oil resins or polyethylene sheeting, helps to prevent some evaporation from the sand surface and stabilizes it so that plants may be established. Fuel-oil mulches have also been used to stabilize dunes but preclude the establishment of natural vegetation. On many stabilized areas, vegetables have been cultivated on experimental production plots, such as the Negev Desert in Israel. These experiments have been successful particularly with the application of nitrogen, phosphorus, potassium, and, occasionally, manure. Increasing effort is being put into land reclamation with these methods, particularly because of the land and food shortages resulting from overpopulation.

Irrigation systems that draw on water from oases have been constructed in some desert areas. The water is distributed by sprinklers and furrows to newly planted vegetation.

Polyethylene greenhouses permit enclosed cultivation in desert areas with no specific preparation of the sand. The environment within the tunnel is controlled and stabilized and normal greenhouse cultivation techniques of temperature and humidity control apply.

CLASSIFICATION of the plant world

Living organisms are placed into one of five kingdoms: Monera—the bacteria; Protista—the algae, diatoms, and dinoflagellates; Fungi—the fungi and lichens; Plantae—the true plants, mosses, ferns, and seed plants; and Animalia—the animals. Each kingdom is divided into one or more divisions, classes, orders, families, and genera. The intent of this table is to show the diversity of some of the major groups of organisms (classes and orders) within the first four kingdoms.

Plant Classification

KINGDOM	DIVISION	COMMON NAME	CLASS
Monera	Schizophyta	cyanobacteria and bacteria	Schizophyceae Schizomycetes Mollicutes
Protista	Chlorophyta	green algae	Chlorophyceae Charophyceae
	Chrysophyta	golden-brown algae yellow-green algae diatoms	Chrysophyceae Xanthophyceae Bacillariophyceae
	Pyrrophyta	dinoflagellates	Desmophyceae Dinophyceae
	Euglenophyta	euglenoids	Euglenophyceae
	Phaeophyta	brown algae	Phaeophyceae
	Rhodophyta	red algae	Rhodophyceae
Fungi	Myxomycota	slime molds	Myxomycetes
	Eumycota	true fungi	Chytridiomycetes Hyphochytridiomycetes Plasmodiophoromycetes Oomycetes Zygomycetes Trichomycetes Ascomycetes Basidiomycetes Deuteromycetes
	Lichenes	lichens	Ascolichenes Basidiolichenes
Plantae	Bryophyta Hepatophyta Anthocerotophyta	mosses liverworts hornworts	Musci Hepaticae Anthocerotae
	Psilophyta Lycophyta Sphenophyta Pterophyta	psilotum club mosses, horsetails, ferns	Psilotopsida Lycopsida Sphenopsida Filicopsida
	Cycadophyta Gnetophyta Ginkophyta Coniferophyta	cycads ephedra ginko conifers	Cycadopsida Gnetopsida Ginkopsida Coniferopsida
	Magnoliophyta	flowering plants	Magnoliopsida Liliopsida

Seed-bearing plants

(SUBDIVISION) CLASS	SUBCLASS	ORDER		
(Gymnospermae) Cycadopsida		Cycadales	Nilssoniales	
Gnetopsida		Welwitschiales	Ephedrales	Gnetales
Ginkopsida		Ginkgoales		
Coniferopsida		Coniferales		
(Angiospermae) Magnoliopsida	Magnoliidae	Magnoliales Piperales	Laurales Aristolochiales	Nymphaeales
	Ranunculidae	Illiciales Ranunculales	Nelumbonales Papaverales	Sarraceniales
	Hamamelidae	Trochodendrales Eupteleales Hamamelidales Urticales Casuarinales	Betulales Myricales Juglandales Cercidiphyllales Didymelales	Eucommiales Barbeyales Fagales Balanopales Leitneriales
	Caryophyllidae	Caryophyllales Polygonales	Cactales Plumbaginales	Theligonales
	Dilleniidae	Dilleniales Theales Passiflorales Begoniales Tamaricales Ericales	Ebenales Malvales Thymeleales Paeoniales Violales Cucurbitales	Capparales Salicales Diapensiales Primulales Euphorbiales
	Rosidae	Saxifragales Fabales Nepenthales Myrtales Rutales Geraniales Cornales	Rhamnales Santalales Proteales Rosales Connarales Podostemales Hippuridales	Sapindales Polygalales Celastrales Oleales Elaeagnales
	Asteridae	Dipsacales Polemoniales Lamiales Calycerales	Gentianales Scrophulariales Campanulales Asterales	
	Liliidae	Triuridales Iridales	Liliales Zingiberales	Orchidales
	Commelinidae	Juncales Bromeliales Eriocaulales	Cyperales Commelinales Restionales	Poales
	Arecidae	Arecales Arales	Cyclanthales Pandanales	Typhales

Glossary

In the following glossary, small capital letters (that is, STAMEN) indicate terms that have their own entries in the glossary.

A

abscission layer A layer of thin-walled cells that forms at the base of a leaf. It breaks at leaf-fall. Similar layers may also occur in bark or at a branch base.

achene A type of fruit consisting of a single dry seed.

actinomorphic Having a radial or star-shaped symmetry, as has the flower of a buttercup.

aerenchyma A type of tissue found particularly in water plants. Generally loose and spongy, it has air spaces between the cells.

akinete A resting cell in some green algae. It is provided with food reserves and serves as a means of VEGETATIVE reproduction.

alkaloids Natural substances found in plants which affect the physiology of animals. Some of these organic bases, such as quinine and strychnine, are used in pharmacology.

alternation of generations The occurrence in the life cycle of a plant of two forms, differing in appearance and number of chromosomes, in regular alternation. One form reproduces sexually, the other asexually.

ameboid Shaped or moving like an ameba, a single cell with a flowing form.

amino acids Fatty acids which contain an amino (NH$_2$) group; these are the building blocks of which PROTEINS are made.

anaerobic respiration A form of respiration in which free oxygen plays no part. Many plants have the ability to break down sugars to alcohols under these conditions. The same process is involved in fermentation of wine and beer.

androecium The collective name for the male parts of a plant, particularly the STAMENS in a flower.

annual A plant that germinates, lives, reproduces, and dies within a single growing season or less time. Many weeds and desert plants have life spans of only a few weeks.

annual ring An annual ring in a cross section of tree trunk is the product of a single year's growth.

annulus Any structure having the form of a ring. The annulus of a fern SPORANGIUM is a ring of thick-walled cells which form a mechanism for tossing out the ripe spores. The ring of cells that ruptures to release spores from a moss capsule is also termed an annulus.

anther The part of the STAMEN of a flower in which the pollen is produced.

antheridium In flowerless plants, the organ or structure in which the male GAMETES are produced.

apical meristem An area of cells at a growing tip which can divide and differentiate into mature tissues.

apocarpous Describes a flower in which the CARPELS are separate from one another.

archegonium The part of a plant that contains the female sex cell. In liverworts, mosses, and ferns the archegonia are small flask-shaped organs with the sex cell in the body of the "flask" and a canal in the neck down which the male cell swims for FERTILIZATION.

ascospore One of the SPORES formed within an ASCUS.

ascus A thin-walled cell in which SPORES form, found in a group of fungi which includes the yeasts and the cup fungi.

association A group of plant species growing together that are characteristic of a particular habitat. The group is often labeled by naming it after the species that is perceived to be the dominant one in the habitat.

autotrophic Describes a plant that manufactures its own food from simple inorganic materials and is not reliant on other organisms for nourishment.

auxin Sometimes called a growth substance, an auxin is any one of a group of substances that act as chemical messengers, being produced in one part of a plant and having an effect at another.

awn A stiff, bristlelike projection, like those which make up the "beard" on the flower spike of barley or wheat.

B

balsams Aromatic compounds produced by conifers.

basidiospore The name for the characteristic SPORES formed by the group of fungi known as Basidiomycetes. This group includes the typical mushrooms and toadstools which bear the SPORES on "GILLS" beneath the cap.

basidium The cell or group of cells from which the SPORES form in Basidiomycete fungi.

biennial A plant that germinates and grows one year, reproducing and dying in the next.

biflagellate Bearing two FLAGELLA, as do some algae.

binary fission A simple cell division in which the NUCLEUS divides into two, followed by a similar division of the cell body.

biosphere That part of the earth and its atmosphere that is capable of supporting living things.

bract A leaflike structure below a flower or flower head. The flower develops in the axil of the bract.

bulb A large underground stem made up of leaf bases which are swollen with food reserves.

bulbil A fleshy bud with food reserves produced above ground and capable of giving rise to a new plant.

C

Calvin cycle Another term for the "DARK REACTION" of photosynthesis, it is named after M. Calvin who received the Nobel Prize in 1961 for his analysis of the process.

calyx The SEPALS, the outer WHORL of flower parts that protects the petals in the bud.

cambium A layer of tissue in a root or shoot that actively divides and gives rise to new permanent tissues. Cambium gives rise to secondary thickening.

canal cell One of the cells in the central part of the "neck" of an ARCHEGONIUM.

capsule A closed boxlike structure, such as the seed case of the poppy flower, or the spore case of a moss.

carbohydrates A group of compounds that includes sugar, starch, and CELLULOSE, consisting of carbon, hydrogen, and oxygen. Oxidation of carbohydrates provides energy for cells. Large-molecule carbohydrates such as CELLULOSE and PECTIN are also important to the structure of plants.

carotene A fat-soluble organic compound that acts as a yellow or orange pigment in plants such as the carrot. Carotin is a spelling variation.

carpel One of the divisions of the female part of a flower, consisting of OVARY, style, and STIGMA. A flower may have a single carpel or several.

cell totipotency The ability of dividing cells to produce differentiated cells, the type depending on the stimuli which affect their growth.

cellulose A large-molecule carbohydrate that forms the cell wall in plants.

chemosynthetic Describes organisms, such as some bacteria, that can synthesize organic material from simple chemicals without the process of PHOTOSYNTHESIS.

chitin A CARBOHYDRATE derivative that forms an important part of the skeleton in invertebrate animals, and is also found in the cell walls of fungi.

chlorophyll The green pigment of plants. It has several types but is chemically always based on a tetrapyrrole ring containing magnesium. Chlorophyll is the substance that traps light energy to begin the process of PHOTOSYNTHESIS.

chloroplast A cell ORGANELLE found in the cells of green plants. It contains the CHLOROPHYLL and forms a unit for PHOTOSYNTHESIS.

chromosome ORGANELLES present in the NUCLEUS that becomes visible by staining at cell divisions. Chromosomes carry the genes that determine the characteristics of a species.

citric acid cycle A process occurring during respiration in the cell. During the cyclic process, which involves the oxidation of pyruvic acid, the compound citric acid is one of the first formed, so giving the name. During this process excess energy from various compounds is converted into high-energy phosphate bonds that store chemical energy for the cell.

cladogram A diagram that attempts to express the relationships of organisms by putting onto the same branch of the diagram only those organisms that share a character that is unique to their group.

climax A stable COMMUNITY of plants that is fully adapted to a particular set of conditions. The climax vegetation may be reached only after a series of stages in which other types of vegetation colonize an area and are subsequently ousted.

coenocytic Describes a fungus that possesses a number of nuclei but has no dividing cell walls.

coleoptile The sheath covering the first pair of true leaves of a grass.

coleorrhiza The sheath covering the young root of a grass.

collenchyma A support tissue found in leaf stalks, midribs, and young stems. It consists of PARENCHYMA cells which may have CHLOROPLASTS but are strengthened by CELLULOSE thickening in the walls.

commensal An organism living with another and sharing its nourishment, but with no ill effect on either.

community A group of plants consistently growing together under a certain set of conditions and forming a recognizable whole, such as the plants characteristic of heathland or oak forest.

companion cell A type of cell found in flowering plants. Companion cells, which retain their nuclei, are associated with the sieve cells of PHLOEM tissue.

compound leaves Leaves made up of several distinct parts or leaflets that are separate down to the midrib, such as the leaves of ash or chestnut.

conidiophore The part of a fungus that bears a CONIDIUM.

conidium A SPORE which forms from a fungal body without any sexual process.

conjugation The joining together of two cells whose nuclei fuse and give rise to new individuals. It occurs particularly in algae, such as Spirogyra.

convergent evolution The development of similar structures or characteristics in two organisms that are not closely related, and believed to be caused by evolution in response to similar conditions in their environments.

cordate Describes a leaf shaped somewhat like the ace of spades in a card deck.

cork A protective layer of dead cells on the outside of a stem or root. Cork cell walls are impregnated with waxy substances that make them relatively impermeable.

corm An underground storage organ consisting of a swollen stem. It may resemble a bulb but has no overlapping fleshy leaves.

corolla The collective name for the petals of a flower.

cotyledon A seed leaf which may remain belowground or may be raised aboveground and become green during GERMINATION, although it is usually a different shape from subsequent leaves.

cross-pollination POLLINATION which involves pollen from one flower traveling to the STIGMA of any other flower of the same species.

cultivar A variety of plant which has been produced in cultivation.

cuticle A waxy layer that covers the outside of epidermal cells.

cutin The mixture of fatty substances that forms the CUTICLE.

cytoplasm The living contents of the cell excluding the NUCLEUS.

D

dark reaction That part of the process of PHOTOSYNTHESIS that does not require light. In the dark reaction, carbon dioxide is fixed and incorporated into compounds in the cell.

daughter cells Cells that are the first generation product of a division of cells.

deciduous Describes trees and shrubs in which the leaves are all shed at a particular season.

decomposers Organisms that help to break down the organic remains of animal and plant bodies into simpler materials.

decurrent Running downwards. This term may be applied to the GILLS of fungi which run down onto the stalk, or to leaves with a continuing wing onto a stem.

dendrochronology The science of analyzing tree rings.

It may be used in calculating the age of trees or lumber, in correlating past events, and in tracing former fluctuations in climate.

dendrogram A diagram which displays data in the form of a branching, treelike pattern, as, for example, some evolutionary trees.

diaspore Any part of a plant that is capable of being dispersed and giving rise to a new plant.

dichotomous Branching into two equal parts, each branch dividing into two again, and so on. This pattern of growth is characteristic of many "primitive" plants. Also the method used to key out species.

dicotyledon A flowering plant in which two seed leaves are present in the seed.

digitate Describes a compound leaf in which the individual leaflets spread like the fingers of a hand.

dimorphic With two distinct forms. Used particularly for species that have a juvenile and an adult form.

dioecious Describes plants in which male and female organs are on different individuals.

diploid The condition of having two basic sets of CHROMOSOMES, one set from each parent, unlike GAMETES, which have a single set.

disaccharides Sugars such as sucrose that are composed of two simple sugars, or MONOSACCHARIDES.

disk florets The small flowers that make up the central part, or disk, of a composite flower such as a daisy. Disk florets are usually trumpet-shaped and symmetrical.

disruptive selection A type of selection in which a species adapts to the surroundings in two or more different ways, which gives an advantage under certain circumstances. In so doing it may abandon the former middle ground of average characteristics for the species, leading to splitting of the species.

dominant A plant species that by its size or number of individuals determines the characteristics of the vegetation in a particular area.

dormant Describes a seed, bulb, or other plant structure that is in an apparently resting condition with no growth and a minimum of metabolism.

drupe A fruit in which the seed is surrounded by a stone and then by a fleshy layer. Plums and peaches are drupes.

E

ectotrophic Finding nourishment from outside, as in some fungi that surround roots and extract nourishment from them.

elaisome An oily particle on the outside of a seed that acts as an attractant to insects such as ants which aid the seed's dispersal.

elater A spirally thickened cell in the spore capsule of liverworts and horsetails. Its movements assist spore dispersal.

embryo-sac The part of a flowering plant that represents the gametophyte generation. It contains eight nuclei including the female GAMETE.

emergent A forest tree that grows so tall that its crown emerges above the rest of the trees.

endoplasmic reticulum The network of folded membranes within a cell that is concerned in making PROTEINS.

endosperm Starchy food tissue that surrounds the embryo in a seed.

endotrophic Feeding from within, as in fungi, which penetrate the root tissues of their host.

endozoochory Transport of a plant seed within an animal.

enzyme A PROTEIN produced by a cell that acts as a catalyst in a living organism, speeding up chemical reactions.

epidermis The layer, usually one cell thick, over the surface of a plant organ (leaf, stem, root, flower).

epiphyte A plant that grows upon another plant, but without being a PARASITE.

F

fertilization The union of a male sex cell with a female sex cell to form a new individual.

fibril A tiny threadlike structure. A root hair.

fix To incorporate into an organism an element from the inorganic surroundings, such as carbon or nitrogen.

flagella Tiny whiplike processes, attached to some single-celled organisms and GAMETES, which by their lashing, move the cells.

floret One of the small flowers in a flower head composite.

follicle A type of fruit formed from a single CARPEL which splits along one side to release its seeds.

forb Any broad-leaved herbaceous plant that is not a grass.

fruiting body A well-defined, spore-bearing structure, such as a mushroom.

funicle The stalk of an OVULE.

G

gametangia Organs that produce sex cells.

gamete A sex cell, capable of taking part in FERTILIZATION.

gemmae Buds or other vegetative outgrowths that are capable of forming new individuals, as in liverworts.

generative cell The smaller of two cells into which a POLLEN GRAIN divides. The generative cell divides to produce two male GAMETES.

geotropism A plant growth response that is initiated by gravity.

germination The breaking of dormancy of a seed or SPORE.

gibberellin A plant hormone that stimulates growth in shoots.

gills The vertical divisions under the cap of a mushroom.

glycolysis The process of breaking down glucose in living tissue to pyruvic acid. This is a stage of respiration that can take place in the absence of oxygen.

granae Minute particles within the CHLOROPLAST of a plant cell in which the CHLOROPHYLL is concentrated.

guard cells The two cells that surround a pore, or stoma, typically on a plant leaf. As they change shape the pore opens or closes.

gynoecium The female part of a flower. Also spelled gynaecium or gynaeceum.

H

halophyte A plant capable of living in very salty conditions, such as on a seashore.

haploid Describes a cell that has only one basic set of CHROMOSOMES. GAMETES are haploid, and some whole plants, such as the gametophyte generation of mosses and liverworts.

haustoria The roots of parasitic plants, or outgrowths from a fungus, that take nourishment from a host's tissues.

heartwood The central core of the hardest wood in a tree.

heath A type of vegetation, found mainly on acid sandy soils, which is dominated by species of the heath family.

hemicellulose A sugar polymer found in some plant cell walls.

herbaceous Describes a plant without a woody stem.

hermaphroditic With both male and female sexual parts on the same plant.

heterosporous Producing two sizes of SPORE, as do some ferns.

heterostyly Having styles of different lengths, as do primrose flowers, thus ensuring CROSS-POLLINATION.

heterotrophic Describes organisms that feed on others

because they are unable to manufacture their own food.

homosporous Producing SPORES all of about the same size.

hydrolysis The splitting of organic compounds through reaction with water.

hydroserel A succession of plant species that begins with the colonization of a wet environment.

hydrostatic skeleton Support that is obtained from the water content of a body.

hydrotropism Plant movements in response to the stimulus of water.

hygromorphic Adapted to a wet habitat.

hyphae The threadlike structures making up the main body of many fungi.

I

indusium A protective cover, particularly the scale or flap that protects the spore-producing bodies in ferns.

inflorescence All the flowers on a stem.

inhibitor A substance that prevents a process from taking place.

innate Originating from within.

integument A covering layer, particularly the outer layer of an OVULE or a seed.

isogamy Sexual reproduction in which the two GAMETES that unite are of similar size, as in some algae.

K

kinin A substance that stimulates growth or division of plant cells.

Krebs cycle A cyclic process in cell respiration. In it pyruvic acid is broken down in the presence of oxygen with the release of energy for cell metabolism.

L

lanceolate Spear-shaped, describing a leaf that is long and tapers to a point at the tip.

lateral bud A bud which forms in the angle between a leaf and a stem, often giving rise to a side-shoot.

latex The milky juice that oozes from some plants when they are damaged. It may contain a complex mixture of compounds. The latex of some plants, such as the rubber tree, is of economic importance.

leached Describes soils in which mineral salts and plant nutrients have been washed out by water percolating down.

leaf-node The point on a stem from which a leaf grows.

legume A plant of the family Leguminosae, which includes the peas, beans, lupines, and acacias. They all have seed pods of the same type.

lenticels Pores through the CORK layer on the outside of a plant. Lenticels permit diffusion of gases in and out.

leptosporangiate A fern in which the SPORANGIUM arises from a single cell.

light reaction The first part of the process of PHOTOSYNTHESIS in which light energy is captured by CHLOROPHYLL and the water molecule is split.

lignified With cell walls thickened with LIGNIN.

lignin A complex compound that becomes mixed with the CELLULOSE layers of a cell wall and gives it strength and rigidity. It gives a plant its "woody" characteristics.

ligule A small flap or scale on the upper surface of a leaf near its base.

M

medullary Belonging to the pith in the middle of a stem.

megaphylls Fern fronds.

megasporangium A structure containing MEGASPORES.

megaspore A large SPORE that gives rise to a PROTHALLUS with female sexual organs.

megasporophyll A leaflike structure bearing MEGASPORANGIA. The scale of a female cone in conifers.

megastrobili The female cones on a conifer tree.

meiosis The process of "reduction division" in which a cell with two basic sets of CHROMOSOMES (the DIPLOID state) divides to produce new cells that contain only one basic set of chromosomes (the HAPLOID state). In animals and flowering plants this reduction takes place when the sex cells form. In lower plants it typically takes place during spore formation when the SPOROPHYTE generation gives rise to the gametophyte generation.

meristem Any region of a plant where the cells are actively dividing, as for example at growing tips.

mesophyll The PARENCHYMA of a leaf, the tissue between the upper and lower EPIDERMIS concerned in PHOTOSYNTHESIS.

microclimate The climate on a small scale as it affects a living organism. It may differ, for example, on two sides of a hill or, on a very small scale, there may be specialized microclimates under a stone or at a tree base.

micropyle A small pore, particularly that in the OVULE of a flower that allows the entry of the tube from a POLLEN GRAIN. This pore persists in the seed and is readily visible in some plants, such as the broad bean.

microsome A small particle within a cell visible by electron microscopy. Microsomes are associated with the ENDOPLASMIC RETICULUM.

microsporangium A structure containing MICROSPORES.

microspore A small spore that gives rise to a PROTHALLUS that bears male sex organs.

microsporophyll The scales of a male cone in conifers. Any leaflike structure bearing MICROSPORANGIA.

microstrobili Male cones of conifers and other gymnosperms.

mitochondria Microscopic bodies within cells that may appear as grains or filaments. They are the "powerhouses" of the cell where energy is released for the cell's work.

mitosis A type of cell division in which both the original cell and those derived from it have two basic sets of CHROMOSOMES (the DIPLOID number), giving a straight-forward replication. This is the normal type of cell division in the nonsexual parts of an organism.

monocotyledon A flowering plant in which there is a single SEED LEAF or COTYLEDON, within the seed, as in grasses or lilies.

monoecious Describes a plant in which both male and female organs are carried on the same individual.

monopodial Describes a stem in which growth is continued from the same growing point at the tip, rather than through the development of strong side branches.

monosaccharides Simple sugars such as glucose and fructose.

moor A vegetation type found on wet peaty soils. According to the degree of saturation of the soil the DOMINANT plants may include sphagnum moss, cotton grass, and heather.

morphogenetic Describes internal secretions that affect the growth and form of a plant.

morphology The structure and form of an organism.

motile Capable of independent movement as a whole organism or sperm cells in more simple plants.

mycelium The web of threads that make up the body of many fungi.

mycorrhiza An association between a fungus and a higher plant with apparent benefits for both. The MYCELIUM of the fungus forms an intimate association with the plant roots.

N

nectar-guides Markings on flower petals that guide insects to the nectar and help to ensure POLLINATION. The marks may sometimes only be visible to insects with vision extending into the ultraviolet range.

nucellus The body of the OVULE of a plant; the thin-walled cells between the embryo and its INTEGUMENT.

nuclear membrane The thin membrane surrounding the NUCLEUS of a cell.

nucleic acids Complex organic acids present in the NUCLEUS of cells and, to a lesser extent, in surrounding CYTOPLASM. Nucleic acids provide a coded pattern for PROTEIN synthesis in the cell and, in the CHROMOSOMES, carry genetic information from one generation to another.

nucleolus ORGANELLES within the NUCLEUS where ribosomal RNA's are synthesized.

nucleus The major ORGANELLE of the cell, usually shaped as a sphere or spheroid. It controls the synthetic reactions in the cell and contains the CHROMOSOMES that are responsible for heredity.

nyctinasty Movements, sometimes called sleep movements, shown by plants in response to the alternation of day and night, as the opening and closing of tulip flowers.

O

oogamy A type of sexual reproduction in which there is a union of a large nonmotile female egg-cell with a smaller MOTILE male GAMETE, as in algae such as Volvox.

oogonia Organs that bear female reproductive cells in algae.

oospheres Unfertilized female egg cells of algae or higher plants.

organelles A general term for the distinct units that are visible within cells by microscopy, such as the NUCLEUS or the CHLOROPLASTS. Each type of organelle has its own function.

osmosis The process by which water diffuses through a cell membrane permeable to water from a weaker solution to a stronger solution.

osmotic pressure A measure of the concentration of a solution. It is the pressure which would need to be applied to prevent the entry of water into the solution by osmosis.

ova Female egg cells.

ovary The part of the flower that contains OVULES, at the base of the CARPEL.

ovate Egg-shaped, describing leaves wider near the base.

ovule In a flowering plant the structure which contains and surrounds the female egg-cell. After fertilization it develops into the seed.

P

palmate Describes leaves in which several leaflets arise from the same point and spread like fingers on a hand.

pappus A ring of hairs, such as that on a dandelion fruit, that may aid dispersal.

parasite An organism that lives by taking nourishment from another living organism, giving no benefit in return. Many fungi and some higher plants such as dodder are parasites.

parenchyma A plant tissue consisting usually of thin-walled undifferentiated cells that usually function in water or food storage.

pathogens Disease-producing microorganisms.

pectin A CARBOHYDRATE polymer that cements together plant cell walls.

peltate Describes a leaf or any other plant organ that is more or less circular with a stalk in the middle of the underside.

perennation Survival from one growing season to another.

perennials Plants that survive for several years or more.

perianth The petals and SEPALS that surround the sexual parts of a flower.

pericarp The outer wall of a fruit. Derived from the outer wall of the OVARY, it may be dry or fleshy.

periderm The outer layer of bark, made up of PHELLODERM, PHELLOGEN, and CORK.

petaloid Describes a SEPAL or a leaf that is colorful and resembles a petal.

petiole A leaf stalk.

phellem See CORK.

phelloderm The cells formed to the inner side of the cork cambium, usually of the PARENCHYMA type.

phellogen The cork cambium, the layer of dividing cells running round a stem that produces CORK cells to the outside and PHELLODERM to the inside.

phenotypic The physical appearance of an organism in contrast with genotype.

phloem The plant tissue concerned with the transport of food material in higher plants. Typically the cells of the phloem are long and tubular, with cytoplasmic connections between cells.

photolysis The breakdown of a molecule through the action of light.

photoperiodism The response of a plant to changes in the relative length of day and night.

photosynthesis The process of building up CARBOHYDRATES from carbon dioxide and water that is carried out by green plants. For this they use energy derived from light captured by the CHLOROPHYLL that they contain.

phototactic Describes a movement made in response to the stimulus of light.

phototropism Directional growth of a plant in response to the stimulus of light.

phreatophytes Plants with long roots penetrating deep down into the soil for water.

physiological Relating to life processes.

phytoplankton Plants, especially microscopic algae, that drift in the water that surrounds them.

pinnate Describes a compound leaf consisting of two rows of leaflets, one on each side of the central stalk.

pioneer species The first plant species to colonize a newly available patch of ground.

pistil The female parts of a flower; the GYNOECIUM.

plantlet A small young plant. Refers particularly to plants that reproduce vegetatively.

plasma membrane The outer membrane of a cell surrounding the living cell contents.

plasmodium A type of organization shown by slime molds in which an AMEBOID mass with many NUCLEI moves and digests organic matter.

plastids Microscopic ORGANELLES found in the CYTOPLASM of plant cells. Many plastids contain pigment, for example the green CHLOROPLASTS.

plumule The growing point of the first shoot of a plant, present in the embryo within the seed.

pneumatophores Specialized roots that grow up from water in some plants, such as swamp cypress. Sometimes used to describe air bladders in water plants.

podsolized Turned into a podsol, a layered soil in which plant nutrients have been leached from the upper layers.

pollen grains MICROSPORES that produce the male sex cells of a flowering plant or conifer. Pollen grains contain NUCLEI with one basic set of CHROMOSOMES (the HAPLOID number).

pollen mother cells The cells within the ANTHERS that each divide to produce four POLLEN GRAINS.

pollen sacs Cavities in the ANTHER in which pollen forms.

pollination Any process by which ripe pollen is trans-

ferred from the ANTHER to the STIGMA of a flower.
pollinators Animals that assist in POLLINATION.
polygamous In botany, describes a species, such as the ash, that bears male, female, and HERMAPHRODITIC flowers.
pome A type of fruit in which the outer flesh is formed from the RECEPTACLE. Apples and pears are pomes.
primary root The first root that develops in a plant, derived from the RADICLE present in the embryo.
prokaryotes Organisms such as bacteria and cyanobacteria (blue-green algae) that lack a distinct NUCLEUS within their cells.
propagation The increase of plant numbers by vegetative means. In cultivation this increase is achieved by such devices as taking cuttings.
proteins Essential constituents of living cells, proteins are compounds with large molecules built up of AMINO ACIDS.
prothallus The small plant formed when a SPORE germinates and contains the organs that produce the sex cells, as in the life cycles of ferns. Sometimes refers to the initial growth stages of lichens.
protonema The threadlike plant that first develops from the asexual SPORE of a moss, giving rise later to the typical moss plant, or plants.
pulvinus A thick, fleshy base to a leaf stalk.

R

radially symmetrical Describes an organism in which like parts are arranged around a vertical axis. Flowers such as buttercups and wild roses are radially symmetrical.
radicle The embryo root contained within a seed.
receptacle The upper end of a stalk bearing the parts of a flower. In some algae it is the end of a branch bearing reproductive organs. In some other plants it is a cuplike structure bearing reproductive organs.
reduction division See MEIOSIS.
reticulate With a netlike appearance.
rhizoids Small rootlike projections that anchor a plant to a surface. They may be present in mosses, liverworts, algae, and ferns.
rhizomes Underground stems, usually horizontal, persisting for more than one growing season. They may be rootlike, but have buds and leaves. Irises and many grasses are examples of plants with rhizomes.
rhizophores Special root-bearing branches in club mosses.
root-hairs Hairlike outgrowths from plant roots that increase the absorptive area.

S

samara A single-seeded type of fruit in which there is a membrane that forms a wing which assists in dispersal, as in maple and ash seeds.
saprophyte A plant that gets its food from dead and decaying organic matter rather than making its own.
scalariform Ladderlike in appearance, as are some types of cell thickening, or some forms of conjugation in algae.
sclereid A cell with tough lignified walls, usually with no living contents. "Stone cells" of this type give the gritty texture found in pears, but are found in many other plants in which toughness is required.
sclerenchyma Tissue made up of SCLEREIDS, the supporting tissue of a plant, especially in herbaceous stems.
sclerophyllous Describes plants with hard, tough leaves that are often small and have a thick CUTICLE. Such plants are characteristic of dry places.
secondary meristem A region of cell division in a plant that produces an increase in thickness rather than elongation of the stem or root. The CAMBIUM and CORK CAMBIUM are secondary meristems.
seed leaf A COTYLEDON, a thick, fleshy storage leaf found in the plant embryo inside the seed.
self-pollination Pollination of a STIGMA by pollen from the ANTHERS of the same flower.
self-sterile Describes the condition in which a flower is unable to be fertilized by its own pollen.
semipermeable membrane A membrane that allows the solvent of a solution to pass through, but does not allow the passage of the dissolved substance with its larger molecules. In living systems the solvent is usually water.
sepals The outer ring of flower parts, usually green and leaflike, but sometimes colored like petals, as in iris and gladiolus.
septate Divided by walls into compartments.
sessile With no stalk, as in some leaves or flowers. Fixed in position, rather than mobile.
seta Any bristlelike structure. The stalk that bears the spore capsule in mosses and liverworts.
sexual reproduction Reproduction in which a male and female sex cell fuse to produce a new individual.
sieve tube A cell in the PHLOEM that transports food materials.
sorus An associated group of spore-bearing organs (SPORANGIA) in ferns. Also refers to groups of sporangia in other lower plants.
spermatophyte A seed-bearing plant, gymnosperm or angiosperm.
spermatozoids Male sex cells which can move by means of FLAGELLA, as in ferns, liverworts, and mosses.
sporangia Walled structures in which SPORES are produced in ferns, or equivalent parts in other plant groups.
sporangiophores Stalked structures bearing SPORANGIA.
spore A specialized reproductive cell that can give rise to a new plant.
sporophylls Leaflike structures that bear SPORANGIA.
sporophyte In the ALTERNATION OF GENERATIONS found in plants, the generation with two basic sets of CHROMOSOMES (the DIPLOID number) that reproduces asexually by means of SPORES.
stabilizing selection A type of natural selection in which the same characteristics prove successful over many generations, minimizing the changes in the organism.
stamen The male reproductive organ of a flower, consisting of a stalk bearing an ANTHER within which the male sex cells (pollen) are produced.
stigma Part of the female reproductive organ of a flower, the stigma is the swollen tip of the stalk (style) on which pollen sticks, leading up from the OVARY.
stolons Creeping horizontal stems of a plant, from which new plants may arise, radiating from a central rosette. The term is sometimes confined to the description of such stems that grow underground to distinguish them from above-ground runners.
stomata The pores in leaves and young stems through which gas exchange can take place. Each stoma is surrounded by a pair of GUARD CELLS.
stomium The part of the spore case of a fern that breaks open to release SPORES.

T

tendril A modified leaf or stem, threadlike and used for climbing, as in peas.
transduction A process by which genetic material is transferred from one bacterium to another by means of a phage.
translocated Describes materials that are moved from one part of a plant to another.
transpiration The process in which leaves give off large quantities of water vapor into the air. The water passes through the plant from roots to leaves in what is called the transpiration stream.

tropism A growth movement in response to an external stimulus.

tube nucleus One of two NUCLEI in the pollen tube or MICROSPORE, it plays a part in regulating the growth of the pollen tube and disintegrates before FERTILIZATION.

tubers Swollen underground stems containing food stores, as in potatoes.

turgor pressure Rigidity of plant tissue due to pressure of water within the cells stretching the cell walls.

tyloses Intrusions from other cells that may eventually block XYLEM vessels.

U

unicellular Consisting of a single cell.

unilocular With a single compartment, as have some OVARIES.

V

vacuoles Cavities in the substance of a cell, generally fluid-filled.

vascular bundle One unit of the fluid-conducting tissue of a plant, consisting of groups of XYLEM and PHLOEM cells.

vascular plant Those plants that have a specialized fluid conducting system, the vascular system.

vegetative Describes functions that are carried out without sexual reproduction being involved.

vessels Water-conducting tubes in the XYLEM made up of cells joined end to end.

volva A ring of tissue on the stalk of a ripe mushroom.

W

whorl A single ring of leaves, petals, or other organs.

X

xanthophyll One of a group of yellow or orange hydrocarbons that act as pigments in plants.

xeromorphic With a structure adapted to dry conditions. This term is usually used to describe leaves.

xerophyte A plant that is able to survive in very dry conditions.

xylem The water-conducting tissue of plants, made up of two types of cell, VESSELS and tracheids. It forms the wood tissue and so also provides mechanical support.

Z

zoochory The dispersal by animals of seeds or SPORES.

zoophytes Plantlike animals.

zygomorphic With only one plane of symmetry. Bilaterally symmetrical, as the flower of a sweet pea.

zygospore A thick-walled SPORE. Such spores form a resting stage in some algae and fungi.

Index

A

abscission layer, 25, 75, 146
acacia, 66, 103, 128, 132, 137, 142
acanthus (Acanthaceae), 92
acetate rayon, 120-121
achene, 146
Achras zapota, 133
acid rain, 141
actinomorphic, 146
adaptations, plant, 78-109
    aquatic plants, 104
    desert, 98
    freshwater plants, 106-107
    grassland, 98
adder's tongue, 38
adenosine diphosphate (ADP), 22
adenosine triphosphate (ATP), 21, 22
aerenchyma, 146
Africa
    Congo Basin, 90
    Ruwenzori mountains, 82, 83
    savanna, 102, 103
agronomists, 8
akinete, 146
alcohol, 127
alder, 49, 77, 81, 84, 109, 139, 140-141
alfalfa, 142
algae farming, 136
algae (Phyophyta), 24, 28-29, 104-105, 106
Alismataceae (water plantain), 57
alkaloids, 23, 128, 146
aloe, 57
alpine snowbell, 83
alternation of generations, 17, 34-35, 36, 37, 146
Amazon Basin
    forests, 90, 94-95
Amazonian waterlily, 78
amber, 133
ameboid, 146
amino acid, 146
amylopectin, 22-23, 23
amylose, 22-23
Andreaea, 85
androecium, 146
anemone, 48, 89
angiosperms, 13, 46-77 (*see also* Flowering plants)
animals, 17
    grazing, 135
    and plant reproduction, 47, 77, 92
annual plants, 146
    desert, 98-99
    herbaceous, 63
annulus, 146
ant
    seed dispersal, 52

anther, 47, 48, 146
antheridium, 35, 37, 39, 146
Anthoceros, 35
Anthocerotophyta (*see* Hornwort)
antibiotics
    from fungi, 33
Aplanatae (bread mold), 29
apocarpous, 146
apple, 50, 77
aquatic plants, 58, 104-107
araucaria, 43
arboriculturist, 8
archegonium, 35, 37, 39, 146
Archichlamydeae, 60-61
Arctic lupine, 85
Arctic tundra, 84-85
Argentina
    land reclamation, 142
Arisaema triphyllum, 51
arrow-head plant, 57, 106
arum, 57, 64
Ascomycetes, 31-32, 108
ascospore, 146
ascus, 146
ash, 53, 75, 89, 112, 138, 139
aspen, 87
Aspergillus, 32
aspirin, 111
association species, 89, 146
ATP (*see* Adenosine triphosphate)
atropine, 23, 128, 134
Australasian cycad, 40
Australia
    grasslands, 101, 102, 103
Australian acacia, 128
Australian mountain ash, 70
Austrotaxus, 45
autotrophic plant, 12, 146
autumn crocus, 129
auxin, 24, 70, 146
awn, 146

B

Bacillariophyta (diatoms), 27, 29
Bacilli, 26
bacteria (Schizophyta), 12, 26-28, 108
balata, 132-133
balsa, 110, 112
balsam firs, 86
balsam pear, 69
balsams, 146
bamboo, 59, 119
banana, 57, 58
baobab, 102
bark, 73-74
    savanna trees, 102
    source of drugs, 128
barley, 59, 126
Basidiomycetes, 32-33, 108
basidiospore, 146

basidium, 32-33, 146
basil, 63
basswood, 70
bast fibers, 124
bat
    and plant reproduction, 52, 58, 76
beans (Leguminosae), 61 (*see also* Leguminosae)
beech, 89, 138, 139
bee pollination, 51
beer, 128
belladona, 23, 128, 134
bent grass, 100
biennial plants, 62-63, 146
biflagellate, 146
bignonia (Bignoniaceae), 68
bilberry, 81, 85, 87
binary fission, 13, 27, 146
bindweed, 65
biosphere, 146
birch, 60, 74, 81, 82, 84, 86, 89, 113, 141
birch forest, 86
bird
    and plant reproduction, 52, 58, 76
bird-of-paradise flower, 56
bird's nest fern, 78-79
bird's nest orchid, 64
bladderwort, 64, 107
blockboard, 113
bluebell, 49, 61, 89
blue grama grass, 102
bluegrass, 100, 101, 102
blue-green algae (*see* Cyanobacteria)
bluestem, 101
bog arum, 58
bog myrtle, 96
bogs, acid, 81
bonsai trees, 71
botanists, 8
bottle tree, 90, 102
bracken fern, 38, 55
bracket (*see* Basidiomycetes)
bracket fungus, 33
bract, 146
branches, 13
breadfruit plant, 93
bread mold (Aplanatae), 31
brome, 102
bromeliads (Bromeliaceae), 56, 62, 93
Brophyllum, 54
broom, 66, 142
broomrape, 65, 96
brown algae (Phaeophyta), 29
bryony, 65
Bryophyta (*see* Liverwort; Mosses)
buckeye, 75
buds, 13

buffalo grass, 54, 102
buildings, 110
bulbil, 54, 147
bulbs (plants), 55, 147
    herbaceous plants, 64
bully tree, 132-133
bulrush, 81, 106
burdock, 52
bur reed, 107
Butomaceae (rushes), 59
butterbur, 65
buttercup, 23, 48, 60, 82
butterwort, 81

C

cabbage (Cruciferae), 61, 63
cacao tree, 93
Calabrian primrose, 78
calendaring, 119
California
    Death Valley, 98
California nutmeg, 45
Calvin cycle (*see* Dark reaction)
calyx, 147
cambium tissue, 18, 19, 147
    dicotyledons, 60
    growth rings, 71
    trees, 70-71
camellia, 66
camphor tree, 133
Canada
    fens, 81
    forests, 86-87
    prairie, 101-102
Canadian pondweed, 54, 58, 106, 107
canal cell, 147
cancer
    drugs from plants, 95
canola, 111
capsule, 59, 147
caraway, 63
carbohydrase, 23
carbohydrates, 23, 147
carbon dioxide
    and deforestation, 136
    freshwater plants, 107
    in photosynthesis, 20
carbon-oxygen cycle, 94
cardboard, 118-119
careers
    botany, 8-11
Caribbean pine, 138
carnivorous plants, 61, 64-65, 79, 80, 81, 107
carotene, 21, 147
carpel, 48-49, 147
carrot, 63, 64
cartwheel flower, 65
castor, 127, 128, 129
catkins, 49, 50, 76, 77

Index

celandine, 64
cell division, 27
cellophane, 121
cells
    animal, 12
    bacteria, 27
    plant, 12, 18
cellulose, 18, 22, 111, 120-121, 147
Cenozoic period
    flowering plants, 46
Cephalotaxaceae (plum yew), 43
cereals, 59
chalk
    flagellates, 27
chamise, 97
chaparral, 67, 97, 99
chemicals
    plant response to, 23
    from wood pulp, 129
chemosynthetic bacteria, 147
cherry, 50
chestnut, 73, 128, 141
chewing gum, 133
chick peas, 63
chicle, 133
Chinese cherry, 73
chipboard, 114
chitin, 147
Chlamydomonas, 29
Chlorobium bacteria, 26
chlorophyll, 20-21, 88, 147
Chlorophyta (*see* Green algae)
chloroplasts, 18, 147
chromosome, 147
    bacteria, 27
Chrysanthemum, 48, 62
Chrysophyta (golden-brown algae), 29 (*see also* Diatoms; Golden-brown algae)
cinchona tree, 128, 134
circumnutation, 147
citric acid cycle (*see* Krebs cycle)
cladograms, 15, 147
class, 14
classification, 16
    flowering plants, 77
    plants, 14-15
    table, 144-145
clear-cutting, 94, 139, 141
Clematis, 52
climate
    conservation, 136
climax community, 147
climbing plants, 65, 68-69
    in rain forest, 90, 93
Clostridium bacteria, 27
cloud forest, 90, 91
club moss (Lycopsida), 13, 36-37, 79, 82
coating pigments, 119
coca, 128
cocaine, 128
cocklebur, 25
coco-de-mer, 53
coconut palm, 53, 59, 77
codeine, 134
coenocytic plant, 147
colchicine, 23
coleoptile, 147
coleorrhiza, 147
collenchyma tissue, 19, 63, 147
colonization of plants, 67
Colophospermum mepane, 103
commensal organism, 147
Commiphora, 103
companion cell, 147
Compositae (lettuce), 61, 78
compound leaf, 147
Congo Basin
    forests, 90
conidiophore, 147
Coniferophyta (*see* Conifer)
conifer (Coniferales), 13, 43-45, 89
Coniferopsida (*see* Conifer; Ginkgo; Yew)
conjugation, 13, 25, 147
Conocephalum, 34
conservation, 134-136
coppicing systems, 140-141

coral root orchid, 64
cordate leave, 147
cord grass, 80, 104-105
coriander, 63
cork cambium (phellogen), 18, 73, 74, 150
cork oak, 74, 115
cork (product), 110, 115
corm, 55, 57, 147
corn, 20, 59, 135
corolla (plant), 147
cotton, 111, 122-123
cottongrass, 81, 85
cotyledon, 56-61, 59, 147
couch grass, 142
cowslip, 55
crambe, 127
cranberry, 81, 85
crassulacean acid metabolism, 98
creeping jenny, 62
Cretaceous period
    flowering plants, 46
    monocotyledons, 56
    seed ferns, 40
crocus, 49, 55, 59, 129
cross-pollination, 50-51, 147
crowberry, 85
crow garlic, 54
Cruciferae (cabbage), 61
cucumber (Cucurbitaceae), 61, 68
Cucurbitaceae, 61, 68
cultivar, 147
cup fungi, 31-32
cuprammonium rayon, 120
Cupressaceae (cypress), 43
curare, 77
cuticle, 105, 147
cutin, 147
cyanobacteria, 12, 13, 28, 108
Cyatheaceae (tree fern), 37
cycad, 13, 40-41
Cycadophyta (*see* Cycad)
Cycadopsida (*see* Cycad; Seedfern)
Cycas, 40
Cyclamen, 52
Cynorchis, 52
Cyperaceae (sedge), 53, 59, 81, 85
cypress (Cupressaceae), 43
cytoplasm, 18, 147

D

daffodil, 59
dahlia, 55, 62
daisy, 61, 78
dandelion, 50, 52, 62, 64, 133
Danthonia frigida, 101
dark reaction, 21, 147
date palm, 56, 57
Datura, 52
Death Valley, 98
deciduous forests, 88-89
deciduous plants, 147
    saltwater marshes, 80
    trees, 74
decomposers, 13, 147
deforestation, 94, 136
dendrochronology, 71, 147-148
dendrograms, 15, 148
deoxyribonucleic acid (*see* DNA)
desertification, 143
desert plants, 78, 98-99
Deuteromycetes, 32
Devonian period
    appearance of ferns, 38
diaspore, 148
diatoms (Bacillariophyta), 27, 29
dichogamous plant, 148
dichotomous growth, 148
Dicksoniaceae (tree fern), 37
dicotyledons (Magnoliopsida), 60-61
digitalis (drug), 128
Digitaria, 103
digitate leaf, 148
dill, 25, 63
dimorphic flower, 149
dinoflagellates (Pyrrhophyta), 29
dioecious plants, 148

ephedrales, 43
    yew, 45
Dioscoreaceae (yam), 57, 59
diploid cell, 17, 148
directional selection, 79
disaccharides, 22, 148
diseases
    caused by bacteria, 26
disk florets, 148
disruptive selection, 79
division, 14
DNA (deoxyribonucleic acid)
    bacteria, 27-28
    cyanobacteria, 12
dodder, 61, 79, 96, 97
dominant species, 148
dormancy, 85, 148
    desert plants, 97
    grasses, 100
Douglas fir, 113, 139
dracena, 70
dragon tree, 57
Drimys, 72
drought-deciduous plants, 99
drugs, 95, 128, 134
drupe, 59, 148
duckweed, 58, 62, 105, 107
duramen (heartwood), 72-73, 148
durian, 52, 92
dye plants, 73, 128-129
dynamite, 29

E

earth star fungus, 33
ebony, 73
economic value
    algae, 29
    cycads, 41
    fungi, 33
    rain forest, 93, 94-95
ectotrophic nourishment, 148
edelweiss, 83
eelgrass, 58, 104
elaisome, 148
elater, 148
electricity, static, 133
elm, 77, 89
embryo-sac, 148
emergent layer, 90, 148
empoldering, 136-137
Encehalartos latifrondia, 41
endoplasmic reticulum, 18, 148
endosperm, 58-59, 148
endotropic mycorrhiza, 96, 148
endozoochory, 148 (*see also* Animals)
English Channel
    chalk cliffs, 27
environmental conditions, 80-107
enzyme
    in plants, 23
    in seeds, 25
Ephedrales, 41, 43
ephemeral plants, 148
epidermis, plant, 18, 148
epiphyte, 78-79, 86, 148
    bromiliads, 56
    ferns, 38
    in rain forest, 90, 93
erosion, 141
Escherichia bacteria, 27
esparto, 112
ethyl alcohol, 127
eucalyptus, 66, 70, 89, 103, 137, 139, 142
Euglena, 13, 27
Euglenophyta (*see Euglena*)
eukaryotes, 12
Eumycota (*see* Fungi)
Euphorbia, 52
Europe
    forests, 88-89, 139
    grasslands, 100
evergreen forests, 89
    rain forest, 90
evergreen plants, 86
evolution
    flowering plants, 46-47

mosses and liverworts, 35
plants, 15, 28, 79
seed ferns, 40
explosives, 121
eyebright, 97

F

"fairy rings," 33
family (classification), 14
fats in plants, 23
fens, 81
fermentation (*see* Respiration, anaerobic)
ferns (Filicopsida), 13, 38-39, 79
fertilization
    of plants, 51, 148
fescue, 100, 101, 102
fiber products, 122-123
fibril, 148
Filicopsida (*see* Ferns)
fir, 86, 112
fir club moss, 82
fire
    forest, 139
    forest management, 140
    grasslands, 101
    heaths, 96-97
    savanna, 102-103
fire fighting
    foresters, 9
fish foods
    algae, 29
flagella, 27, 29, 148
flask fungi, 31-32
flax, 59, 122, 124-125
    linseed oil, 127
floret, 148
floriculturists, 8
flower, 13, 48
    dicotyledons, 60-61
    monocotyledon, 57-58
    in sexual reproduction, 48
    trees, 75-76
flowering plants, 46-77
    tundra, 84-85
follicle, 148
food crops, 135-136
    dicotyledons, 61
    herbs, 63
forb, 101, 148
foresters, 9
forest management, 138-141
    foresters, 9
forests
    affecting climate, 136
    endangered, 135
forests, coniferous, 86-87
forests, temperate, 88-89
forests, tropical, 90-93 (*see also* Rain forest)
fossil plants, 15
    seed ferns, 40
foxglove, 63, 128
frangipani, 47
freesias, 55
French cotton plant, 132
frogbit, 107
frost
    plant damage, 107
fructose, 126
fruiting body, 148
fruits, 50, 51
fruit trees, 76-77
fungal hyphae
    conifers, 45
fungi (Mycota), 13, 17, 30-33
    endotropic mycorrhiza, 96
    in forest floor, 86
    saprophytic, 108-109
    symbiotic, 58-59
    as tree diseases, 139
funicle, 148
furze, 96, 97

G

gametangia, 148
gamete, 148

gametophyte generation, 17, 34
garlic, 59
gemmae, 35, 148
generative cell, 148
gentian, 61
genus, 14
geotropism, 23
geranium, 64
germination, 148
  desert plants, 98-99
  monocotyledons, 58-59
ghost orchid, 64
gibberellin, 24, 25, 148
gills (plant), 148
ginger (Zingeberaceae), 92
Ginkgophyta (see Ginkgo)
ginkgo, 43
ginseng, 128
gladiolus, 49, 58, 59
glasswort, 78, 80, 81, 105
global warming (see Greenhouse effect)
glucose, 22, 126
  in plant respiration, 21-22
glycerol, 23
glycolysis, 21, 22, 148
Gmelina arborea, 138
Gnetales, 43
Gnetopsida, 42-43
golden-brown algae (Chrysophyta), 29
goldenrod, 63
Golgi bodies, 26
Gonyaulax, 29
Goodyear, Charles, 130
gorse, 66
grain, 50
Graminaceae (grasses), 56, 64, 100
gram-positive bacteria, 27
granae (thylakoids), 20, 148
grasses (Graminaceae), 56, 64, 100
grasslands, 100-102
gravity
  plant response to, 23, 24
grazing, 135, 143
Great Lakes
  forests, 87
green alder, 84
green algae (Chlorophyta), 29
greenhouse effect, 94, 136
groundsel, 63, 82
growth
  trees, 70-71
guard cells, 148
gum, 132, 133
gum arabic, 132, 133
gum benzoin, 133
gum tree, 142
gunnera, 65
gutta-percha, 132
gymnosperms (see Conifer; Cycads)

H

halophytes, 78, 80-81
haploid cell, 17, 148
hardboard, 115
hardwood, 61, 71, 112
haustoria, 148
hazel, 48, 89, 141
heartwood (duramen), 72-73, 148
heath, 96-97
  saprophytes, 108
heather, 61, 66, 81, 96, 97
hellebore, 64
hemicellulose, 148
hemlock, 87, 89, 139
hemp, 59, 124, 125
henna, 129
Hepatophyta (see Liverwort)
herbaceous plants, 62-65, 148
herbals, 15
hermaphroditic plants, 47, 48, 77, 148
heterosporous, 148
heterostyly, 50
heterotrophic plants, 148-149
Himalayan balsam, 65

hollyhock, 63
homosporous fern, 39, 149
honeysuckle, 69
Hooke, Robert, 18
hop, 69
hornbeam, 141
hornwort, 107
horsetail (Sphenopsida), 13, 37
horticultural scientists, 9
horticultural therapists, 10
hybridization, 135
Hydra, 108
hydrogen carrier, 21
hydrolysis, 149
hydrophytes (see Aquatic plants)
hydroseral succession, 149
hydrostatic skeleton, 19, 149
hydrotropism, 25
hygromorphic, 149
Hyparrhenia, 103
hyphae, 149

I

ilareta, 82, 132
India
  forests, 90
Indian almond, 91
Indian grass, 101
Indian pipe orchid, 64, 108
indigo, 128
indusium, 149
inflorescence, 48, 61, 149
inhibitor, 149
insect-eating plants (see Carnivorous plants)
insecticides, 139
insect pollination, 47, 50-51
  desert plants, 99
  trees, 76
insulation
  rubber, 130
  wood products, 110
integument, 149
International Code of Botanical Nomenclature, 14
inulin, 23
Iridiris, 50, 55, 59, 64
iris, 57
irrigation, 143
Isoberlinia, 103
Isoetales (quillwort), 36-37, 107
isogamy, 149
Israel
  Negev Desert, 143
ivy, 69

J

jack-in-the-pulpit (*Arisaema triphyllum*), 51
jojoba, 127, 135
Joshua tree, 70
June grass, 102
jungle plants, 68
juniper, 87
jute, 124, 125

K

kangaroo grass, 103
Kansas
  Niobrara Chalk Formations, 27
kapok (silk-cotton tree), 77, 91
kelp, 105
kenaf, 125
Kentucky bluegrass, 55
kiln drying, 112
kinin, 149
kraft pulping, 117
Krebs cycle, 21-22, 147, 149

L

Labrador tea, 81
laminated lumber, 113
lanceolate, 149
landscape architects, 10
Languncularia racemosa, 81

larch, 43, 81, 86, 138
lateral bud, 149
latex, 132-133, 149
laver bread, 105
leaf fall, 88-89
leaf-node, 149
leatherleaf, 81
leaves
  conifers, 44
  dicotyledons, 60, 61
  heath plants, 96
  herbaceous plants, 64
  loss of, 75
  monocotyledons, 57
leeks, 59
legumes (Papilionaceae), 50, 65, 109, 149
Leguminosae, 61
lenticels, 149
lentil, 63
leptosporangiate fern, 149
lettuce (Compositae), 61, 78
liana, 93
lichens, 13, 108-109
  litmus tests, 129
  tundra, 85
light
  and photosynthesis, 21
  plant response to, 24, 25
light reaction, 21, 149
lignin, 63, 70, 116, 129, 149
lignum vitae, 110
ligule, 149
lily family, 57, 60
Linnaeus, Carolus, 14, 16
liverwort, 13, 34-35, 82
lobelia, 82
logania, 77
logging, 112, 139-140
logwood, 73
lousewort, 84
lumber, 110-111
Lunularia, 34
lupine, 53, 54, 64, 85, 142
Lycopodiales (see Club moss)
Lycopsida (see Club moss; Quillwort)
Lyginopteris, 40
lysosomes, 18

M

madder, 128-129
magnolia, 60, 75, 77
Magnoliophyta (see Flowering plants)
Magnoliopsida (dicotyledon), 60-61
mahogany, 73, 92, 112
maidenhair fern, 38
male fern, 38
maltose, 126
mango, 52
mangrove, 80, 81, 105
mangrove bark, 128
manila hemp, 124
mannan, 22
manzanita, 99
maple, 53, 89, 112, 113, 126
maple syrup, 126
marram grass, 105, 142
marsh, 80-81
  reclaiming, 137
Marsileales (water fern), 39
mat grass, 100
medical botanists, 10
medicinal plants, 95
Mediterranean sweet bay, 66
medullary, 149
megaphylls, 38
megaspore, 149
megasporophyll, 41, 149
megastrobili, 149
meiosis, 149
melon (cucurbitaceae), 61
mercerization, 123-124
meristem, 18, 54, 70, 71, 146, 149, 151
mesophyll, 149

Metachlamydeae, 61
methyl alcohol, 127
Mexican mesquite, 143
Michaelmas daisy, 62
microclimate, 149
micropyle, 149
microsome, 149
microspore, 149
microsporophyll, 41, 149
microstrobili, 149
milk vetch, 133
millet, 59, 103, 142
mining
  landscape destruction, 137
mistletoe, 52
mitochondria, 18, 22, 26, 149
mitosis, 149
Mnium, 34
Monera, 17 (see also Bacteria)
monkey puzzle tree, 43
monocotyledons (Liliopsida), 56-59, 70, 149
monoculture, 135, 139
monoecious plant, 149
monopedical stem, 149
monosaccharides, 149
monsoon forest, 90
Montbretia, 55
Monterey pine, 138, 142
moonwort, 38
moor grass, 97, 100
moorland, 96-97
morel, 31
morning glory, 65
morphine, 23, 134
morphogenetic secretions, 25, 149
morphology, 149
moss, 34
moss campion, 82
mosses (Bryophyta), 13, 34-35, 79
  tundra, 85
motile, 149
mountain
  conifers, 44
mountain arnica, 134
mountain avens, 83
mountain crowfoot, 82
mountain grasslands, 100-101
mountain plants, 82-83
Mucor, 31
multikingdom system, 17
mushrooms, 31-32 (see also Basidiomycetes)
mycelium, 149
Mycophyta (see Slime mold)
mycorrhiza, 109, 149
Myxomycota (see Slime mold)

N

NAD (nicotinamide adenine dinucleotide), 22
nectar, 51
nectar guides, 150
needle-and-thread, 101, 102, 103
Negev Desert, Israel
  land reclamation, 143
Netherlands
  land reclamation, 137
New Guinea
  forests, 90
New Zealand
  grasslands, 100, 102
  land reclamation, 137
nicotinamide adenine dinucleotide (NAD), 22
nicotine, 23, 128
nightshade (Solanaceae), 23, 61
Nitrobacter bacteria, 26
nitrocellulose, 121
nitrogen cycle, 94
nitrogen-fixing bacteria, 26, 96, 109, 142
Nitrosomas bacteria, 26
North America
  forests, 86-87, 89, 139
Nostoc, 29
Novawood, 111
nucellus, 48-49, 150

nuclear membrane, 150
nucleic acid, 150
nucleolus, 150
nucleus (cell)
  plant, 18, 150
nuts, 50
nyctinasty, 25, 150
Nymphaeaceae (waterlilies), 106

O

oak, 70, 74, 75, 89, 112, 115, 128, 138
oarweed, 105
oats, 59, 102
ocean plants, 104-105
  seaweeds/algae, 28, 29
oil, plant, 23
oils, vegetable, 126-127
Olympic National Park, Washington, 95
onions, 55, 59
oogamy, 31, 150
oogonia, 150
Oomyceta (*see* Water mold)
oospheres, 150
opium, 23, 128
orange, 51, 77
orange-peel fungus, 31
orchid, 51, 52, 55, 56, 57, 58, 64, 108
order, 14
organelles, 18, 150
oriental saltwort, 143
Orthotrichum, 34
osmosis
  in plants, 19-20, 150
osmotic pressure, 18, 150
Osmundidae, 38
ova (*see* Ovum)
ovary, plant, 13, 47, 48, 150
ovate leaves, 150
ovule, 48, 150
ovum, 150
oxeye daisy, 48
oxlip, 89
oxygen
  marsh plants, 80
  photosynthesis, 20

P

Palaeotaxus, 45
Palaquium oblongifolia, 132
palmate leaf, 150
palm tree (Palmae), 70
pampas grass, 62, 103, 137
panax, 128
paper, 116-119
Papilionaceae (legumes), 50, 65, 109
pappus, 150
papyrus, 81
parasitic pathogens, 26, 150
parasitic plants, 64, 65, 79, 108-109, 150
parenchyma tissue, 19, 42, 150
parsnips, 63
particle board, 114-115
passionflower, 25, 63, 68
pear, 77
pearlwort, 62
peas, 63
pectin, 22, 150
peltate leaf, 150
penicillin, 32
Penicillium, 32, 33
Pennisetum, 103
Pennsylvanian period
  club mosses, 36
penny-wort, 64
pentose, 22
peony, 52, 62
perennation, 150
perennial plants, 62, 150
perianth, 150
pericarp, 150
periderm, 150
periwinkle, 134
Permian period

gymnosperms, 42
Peronosporales (downy mildew), 31
pests, 139
petaloid leaf, 150
petals, 49
petiole, 150
Phaeophyta (*see* Brown algae)
phellem (*see* Cork)
phelloderm, 19, 150
phellogen (*see* Cork cambium)
phenolic acid, 23
phenotype, 150
pheophytin, 21
Philodendron, 66
phloem tissue, 19
  bast, 124
  dicotyledons, 60
  monocotyledons, 57
  trees, 73-74
photolysis, 150
photomorphogenetic responses, 25
photoperiodism, 25, 150
photosynthesis, 12, 18, 20-21, 150
  in desert plants, 98
phototactic response, 150
phototropism, 69, 150
phreatophytes, 150
Phycophyta (*see* Algae)
Phythium, 31
Phytophthora infestans, 31
phytoplankton, 150
pigments, plant, 128-129
  algae, 29
Pinaceae (*see* Pine)
pine (Pinaceae), 42, 43, 44, 87, 89, 112, 132, 138, 142
pineapple, 56
pine forests, 86, 140
pinnate leaf, 150
Pinophyta (*see* Conifer; Cycad)
pioneer species, 150
pistil, 48, 150
pitcher plant, 64-65, 81
Plantae, 17
plant ecologists, 10
plantlet, 150
plant pathologists, 11
plant products, 110-133, 134-135
  dicotyledons, 61
  monocotyledons, 59
plants
  difference from animals, 12, 17, 20
plasma membrane, 150
plasmodium, 150
plastids, 150
plum, 77
plumule, 150
plum yew (Cephalotaxaceae), 43
plywood, 110, 113
pneumatophore, 150
podocarp, 43, 44
podsol, 96, 150
poisons, 128
polder, 136-137
pollen, 48, 49-51, 150
pollination, 49-50, 150-151
  alpine plants, 83
  conifers, 44
  desert plants, 99
  freshwater plants, 107
  monocotyledons, 58
  trees, 75-76, 77
pollution
  acid rain, 141
  lichen growth, 109
polyamous species, 151
polypody, 39
polysaccharide, 22-23
Polytrichum moss, 34
pome, 151
pomegranate tree, 128
poplar, 49, 52, 73, 138
poppy, 23, 25, 53, 64, 128, 134
potato (Solanaceae), 55
prairie, 101-102
prayer plant, 92

primrose, 50
Primula, 52
Prochloron alga, 28
prokaryotes, 12, 17, 151
propagation (*see* Reproduction, plant; Vegetative reproduction)
prostrate summer cypress, 143
proteins
  in plants, 23, 151
protenema, 151
prothallus, 37, 151
Protista, 17 (*see also* Algae)
protozoa, 17
Prymnesiophytes, 29
Pseudomonas bacteria, 26, 27
Pteridophytes (*see* Club moss; Fern; Horsetail)
Pteridospermales (*see* Seed fern)
puffball (*see* Basidiomycetes)
pulp, wood (*see* Wood pulp)
pulvinus, 151
Pyrrophyta (*see* Dinoflagellates)

Q

quebracho, 128
quillwort (Isoetales), 36-37, 107
quinine, 128, 134

R

radial symmetry, 151
radicle, 151
raffia palm, 124
rag paper, 119
rain forest, 43, 90-93, 94-95
ramie, 123, 124, 125
rat
  seed dispersal, 52
rauwolfia, 128
rayon, 120-121
receptacle, 151
reclamation, 136-137
  shore and desert, 142-143
recycled products
  paper, 116
  wood, 114
red alder, 139
red algae, 29
red currant, 52
red mangrove, 77, 81
redwood, 43, 45, 70, 89, 139
reed, 81, 106
reed grass, 101
reindeer moss, 85, 87
reproduction, plant
  algae, 13, 26-27, 29
  bacteria, 27-28
  club mosses, 37
  cycads, 40-41
  ferns, 38-39
  flowering plants, 48-51
  imperfect fungi, 32
  molds, 31
  mosses and liverworts, 34-35
  mushrooms, 32-33
  vegetative propagation, 54-55
reserpine, 128
resin, 42, 115, 132, 133
respiration
  anaerobic, 22, 146
  plants, 21-22
Rhizobium bacteria, 26, 96, 109
rhizoids, 39, 151
rhizomes, 55, 151
  ferns, 38
  herbaceous plants, 64
rhizophores, 151
rhizophylls, 64
Rhizopus, 31
Rhodomicrobium bacteria, 26
Rhodophyta (*see* Red algae)
rhubarb, 65
ribose, 126
rice, 59, 135
rings, annual growth, 71, 146
river plants, 106-107
rivers, 104-105
  soil erosion, 135

root-nodules, 109
roots, 13
  aquatic plants, 104-105, 107
  climbers, 69
  club mosses, 36
  conifers, 45
  cycads, 41
  dicotyledons, 60
  epiphytes, 78-79, 93
  freshwater plants, 107
  herbaceous plants, 64
  monocotyledons, 57
  savanna plants, 101
  temperate forest plants, 88-89
  trees, 75
roses (Rosaceae), 60, 61, 69, 77
rosewood, 110
rosin, 129, 133
royal fern, 38
rubber, 130-131
rubber dandelion, 133
rubber tree, 130, 132
runners (plants), 55
rush, 59, 81, 106
rust (Uredinales), 33
Ruwenzori mountains, 82, 83
rye, 59, 102, 142
rye grass, 102

S

safflower, 127
saffron, 129
sage, 53
saline environments, 78
salinization, 143
Salmonella, 27
salt-tolerant plants (*see* Halophytes)
saltwater marsh, 80-81
  plants, 105
saltwort, 53, 105, 143
Salvinia, 107
Salviniales (water fern), 39
samara, 50, 151
sandbox tree, 92
sand dune, 142
  reclaiming, 137
sap, 71
saprophytes, 108-109, 151
saprophytic bacteria, 26
sapwood, 71-72
satinwood, 73
savanna, 100-103
saxifrage, 82, 83, 84
scalariform, 151
Scandinavia
  forests, 87
scents in plants, 23, 47, 50-51
Schizophyta (*see* Bacteria)
sclereid cell, 19, 151
sclerenchyma, 63-64, 69, 151
sclerophyllous plants, 151
Scotch broom, 142
screw-pine, 57
scrublands, 67
seablite, 80, 105
sea lavender, 80, 105
sea lettuce, 29, 105
seasoning wood, 112
seaweed, 29, 104-105
sedge (Cyperaceae), 53, 59, 81, 85
seed dispersal, 52-53, 77
  in rain forest, 90-92
seed fern, 40
seed leaf, 151
seeds, 47, 51
  monocotyledons, 58-59
  in rain forest, 90-92
  trees, 76-77
  tundra plants, 84-85
Selaginella, 36, 37
Selaginellales (*see* Club moss)
selection, natural, 79
self-pollination, 49-50, 151
self-sterile plants, 151
semipermeable membrane, 151
sensitive plant, 25
sepals, 49, 151
septate, 151

Index

sessile plant, 151
seta, 151
Setaria, 103
shallot, 59
shellac, 132
shepherd's purse, 63
shrubs, 66-67
    arctic, 84
    in deciduous forests, 89
    desert, 99
    in tundra, 82
Siberia
    forest, 86
sieve tube, 17, 151
silk-cotton tree (kapok), 77, 91
sisal, 59, 123, 124
Sitka spruce, 138
sizing agents, 117
slime mold (Myxomycotina), 30-31
smilax (Smilaceae), 57
smut fungi (Ustilaginales), 33
snow
    and plants, 82-83, 84
snowbell tree, 133
snowdrop, 56, 59
snow tussock, 100
soapberry, 132
softboard, 115
soil
    conservation, 135
    and fungi, 33
    heath, 96
    leached, 149
    rain forests, 93
    salinization, 143
    savanna, 103
soil conservationists, 11
Solanaceae (potato), 23, 55, 61
Solomon's seal plant, 59
sorghum, 59, 103, 142
sorus, 38-39, 151
South America
    grasslands, 102, 103
Southeast Asia
    forests, 90
    land reclamation, 137
soybean, 126-127
special adaptations, 78-109
species, 14
spermatophyte, 151
spermatozoid, 151
Sphagnum moss, 81, 85
Sphenopsida (horsetail), 13, 37
Sphenopteris, 40
spiny durian, 52
spiny sea buckthorn, 142
Spirillum, 26
Spirogyra, 29
sporangiophores, 151
sporangium, 39, 151
spores, 151
    bacteria, 27-28
    bread molds, 31
    ferns, 39
    mosses and liverworts, 35
    slime molds, 30
Sporobolus, 103
sporophyll, 41, 151
    club moss, 36
    cycads, 40-41
sporophyte generation, 17, 34-35, 151
spruce, 42, 86, 87, 112, 113, 138
squirrel
    disperse seeds, 52
squirting cucumber, 53
stabilizing selection, 79, 151
stag's horn club moss, 36
stamens, 49, 151
Stangeria, 40
Staphylococcus, 26, 27
starch
    in plants, 22
stems, 13
    climbing plants, 69
    dicotyledons, 60
    herbaceous plants, 63-64
    monocotyledons, 56
stickseed, 52

stigma, 47, 151
stolon, 151
stomata, 18, 21, 151
    ferns, 38
    heath plants, 96
stomium, 151
stone pine, 44
stoneplants, 99
stonewort, 106
storkbill, 53
strangler fig, 93
strawberry, 55
Streptococci, 26, 27
strip mining, 137
strobilus, 37, 43
strychnine, 23, 77, 128
Strychnos, 77, 128
subclass, 14
succession of species, 86, 87
succulent plants, 98, 99, 105, 142
suckers, 69
sucrose, 126
sugar, 23, 126
sugar cane, 129
sugar cane pulp, 119
sugar maple, 71
sundew, 61, 65, 79, 80, 81
sunflower, 63
sunlight, 20
swamp cypress (Taxodiaceae), 43, 44, 80
swamps, 80-81
    mosses in, 35
sweet flag, 58
sweet grass, 52
switch grass, 101
symbiosis, 108-109
symbiotic fungi, 33, 58-59

T

taiga, 86
tannin, 23, 73, 126, 128
tape grass, 107
taro, 62
Taxales (yew), 45
Taxodiaceae (swamp cypress), 43, 44, 80
Taxol, 95
taxonomists, 8
taxonomy (*see* Classification)
teak tree, 73, 112
tendril, 65, 68-69, 151
Terminalia, 103
terpenes, 23
textiles, 111
    cotton, 111, 122, 123
    hemp, 125
    linen, 122, 124-125
    ramie, 125
    rayon, 120-121
therapists, horticultural, 10
thigmotropism, 24
13-kingdom system, 17
thongweed, 105
Thorp, John, 123
three-leaved rush, 82
thylakoids (granae), 20, 148
thyme, 97
tiger lily, 47
tissues, plant, 18-19
toadstool (*see* Basidiomycetes)
tobacco, 128
tomato, 52
touch-me-not, 53
touch tropism, 69
tracheids, 19
transduction, 151
transpiration, 19-20, 21, 151
    trees, 74-75, 89
transpiration stream, 19
tree fern (Cyatheaceae, Dicksoniaceae), 37, 39
tree lupines, 142
trees, 70-77
    desert, 99
    savanna, 102-103
Triassic Period
    flowering plants, 46

ginkgos, 43
Tricomycetes, 31
triose, 22
Trochodendron, 72
tropical forests, 90-93
tropical shrubs, 66
tropical trees, 71
tropisms, 24-25, 69, 152
truffle, 31
tualang, 90
tube nucleus, 152
tubers, 55, 152
tulip, 25, 55, 59, 60
tulip tree, 77
tumbleweed, 53
tundra
    alpine plants, 82-83
    arctic plants, 84-85
tung, 127
turgor movements, 25, 152
turpentine, 129, 133
tussock, 100
tussock grass, 103
tyloses, 152

U

umbellifers, 64
understory plants, 92-93
unicellular, 152
unilocular, 152
United States
    prairie, 101-102
Uredinales (rust), 33
Ustilaginales (smut fungi), 33

V

vacuole, 18, 152
vanillin, 129
Variolaria, 129
vascular system, 19-20, 152
    aquatic plants, 104
    club mosses, 37
    conifers, 40
    dicotyledons, 60
    herbaceous plants, 63-64
    monocotyledons, 56
    trees, 71-72, 89
vegetative reproduction, 54-55
    climbers, 68
    ferns, 38
    freshwater plants, 107
veneer, 113
Venus's-flytrap, 65, 79
vessels (plant), 152
vetch, 53, 69
vinblastine, 95
vincristine, 95
vines (*see* Climbing plants)
Virginia cowslip, 48
Virginia creeper, 68, 69
virgin's bower, 52
viscose rayon, 120
Vitaceae, 68
volcano
    and aquatic plants, 20
volva, 152
vulcanization, 130-131

W

waferboard, 115
wallflower, 60
walnut, 89, 112
water
    conduction, 19-20
    dispersal of seeds, 47, 53, 77
    movement in trees, 74
    and photosynthesis, 20
    pollination, 51, 107
water buttercup, 106
water-conserving plants (*see* Xerophytic plants)
water cycle, 94
water fern (Marsileales, Salviniales), 39
waterlilies (Nymphaeaceae), 106
water mold (Oomyceta), 31

water plantain (Alismataceae), 57
water soldiers, 107
water starwort, 107
weeds, 62, 63
welwitschia, 42-43
western hemlock, 139
whales, 134-135
wheat, 58, 59, 135
white mangrove, 81
Whittaker, Robert H., 17
wild azalea, 96
willow, 48, 49, 52, 73, 77, 82, 84, 85, 110
willow moss, 106
wind dispersal, 52
wind pollination, 47, 50-51, 76, 77
wine, 127
wintergreen, 87
wisteria, 68
woad, 128
wood, 70, 112-115
wood alcohol, 127
woodchip, 114-115
wood pulp, 111, 116-119
    chemicals, 129
    rayon, 120
wood rays, 71

X

xanthophyll, 21, 152
Xanthoria palietina, 109
xeromorphic plants, 99, 105, 152
    cycads, 41
xerophytic plants, 66, 78, 152
    deserts, 98-99
    Ephedrales, 43
    forests, 86
    heaths, 96
xylan, 22
xylem tissues, 19, 152
    aquatic plants, 104
    climbing plants, 69
    dicotyledons, 60, 61
    monocotyledons, 57
    trees, 72, 74

Y

yam (Dioscoreaceae), 57, 59
yeast, 32
Yeheb bush, 136
yellow horned poppy, 105
yew (Taxales), 45
yucca, 57, 67

Z

Zamia pygmaea, 40
Zingeberaceae (ginger), 92
zoochory, 152 (*see also* Animals)
zoophytes, 151
zygomorphic, 152
Zygomycota (*see* Bread mold)
zygospore, 152

Credits

The following have provided photographs for this book: Cover photo—Tony Stone Images; Michael Abbey/Science Photo Library 24; Aquila Photographics 53; Heather Angel 10, 31, 33, 35, 36, 37, 38, 39, 41, 43, 45, 54, 56, 60, 67, 69, 70, 73, 76, 79, 80, 82, 84, 85, 86, 88, 93, 95, 101, 103, 104, 105, 106, 108, 109, 123, 126, 128, 129, 132, 133, 135, 141; H. Axell/Natural Science Photos 46, 61, 69; C. Banks/Natural Science Photos 65; Biophoto Associates 28, 94, 99; Biophoto Associates/Science Photo Library 27; Frank V. Blackburn/Nature Photographers 141; Lawrence C. Buss 84, 85; Peter Bloomfield 110; Dr Tony Brain/Science Photo Library 14; Robert Brenner/PhotoEdit 10; Paul Brierley 15; Building Research Establishment 19; Brinsley Burbidge/Nature Photographers 25, 65; Dr Jeremy Burgess/Science Photo Library 33, 49, 79; Camerapix/Alan Hutchinson Library 101, 111, 116, 118, 122, 123, 124, 126, 127, 136, 143; W. Cane/Natural Science Photos 62; Kevin Carlson/Nature Photographers 125; Brian Carter 11, 51, 73, 75; Chicago Botanic Garden 9; M. Chinery/Natural Science Photos 99; Ron Croucher/Nature Photographers 138; Timothy Eagan/Woodfin Camp & Assoc. 134; Mary Evans Picture Library 15; M. Freeman/Natural Science Photos 90; Geoscience Features 30, 34, 84, 111; Runk/Schoenberger from Grant Heilman 8, 11; Eric Gravé/Science Photo Library 21, 71; C. Grey-Wilson/Nature Photographers 43, 62; James Hancock/Nature Photographers 140; Robert Harding Picture Library 118, 121; Jan Hinsch/Photo Researchers 16; J. Hobday/Natural Science Photos 91; Holt Studios Ltd 23, 58; Indian Tourist Office 129; Institute of Geological Sciences 133; Nicholas Law 60; Calvin Larsen/Photo Researchers 95; John Lemker/Earth Scenes 83; John Lythgoe/Seaphot Ltd/Planet Earth Pictures 137, 138; G. A. Matthews/Natural Science Photos 88; G. Montalverns/Natural Science Photos 42; Hank Morgan 8; Marion Morrison 78, 93; Tony Morrison 18, 63, 81, 82, 92, 103, 132, 142; Dr Gopal Murti/Science Photo Library 22; Natural Science Photos 76, 85; Nature Photographers 53, 72; G. Newlands/Natural Science Photos 40; Novosti Press Agency 102; The Photosource 47, 74, 112, 143; Rod Planck/Photo Researchers 17; Reed International 117; Juan Renjilo/Earth Scenes 94; R. Revels/Natural Science Photos 23, 107; M. Rose/Natural Science Photos 105; Science Photo Library 123; Shaman Pharmaceuticals 11, 95; Spectrum Dyed Yarns, Inc. 111; Stammers/Greenwood/Science Photo Library 20; L. S. Stepanowicz 29; Tony Stone Images 2, 8, 9, 19, 44, 45, 47, 51, 53, 57, 59, 64, 77, 87, 101, 103, 113, 115, 136; M. W. F. Tweedie/NHPA 92; C. A. Walker/Natural Science Photos 66; P.H. & S.L. Ward/Natural Science Photos 24, 68; Weyroe Ltd 115; WORLD BOOK Illustration by Christabel King